ENTROPY AND ENTROPY GENERATION

Understanding Chemical Reactivity

Volume 18

The titles published in this series are listed at the end of this volume.

Entropy and Entropy Generation

Fundamentals and Applications

edited by

J. S. Shiner

Physiological Institute,
University of Bern,
Bern, Switzerland

KLUWER ACADEMIC PUBLISHERS
DORDRECHT / BOSTON / LONDON

A C.I.P. Catalogue record for this book is available from the Library of Congress.

ISBN 0-7923-4128-7

Published by Kluwer Academic Publishers,
P.O. Box 17, 3300 AA Dordrecht, The Netherlands.

Kluwer Academic Publishers incorporates
the publishing programmes of
D. Reidel, Martinus Nijhoff, Dr W. Junk and MTP Press.

Sold and distributed in the U.S.A. and Canada
by Kluwer Academic Publishers,
101 Philip Drive, Norwell, MA 02061, U.S.A.

In all other countries, sold and distributed
by Kluwer Academic Publishers Group,
P.O. Box 322, 3300 AH Dordrecht, The Netherlands.

Printed on acid-free paper

Table of Contents

PREFACE

Entropy and entropy generation are fundamental quantities. They play essential roles in our understanding of many diverse phenomena ranging from cosmology to biology. Their importance is manifest in areas of immediate practical interest such as engineering as well as in others of a more fundamental flavor such as the origins of macroscopic irreversibility from microscopic reversibility and the source of order and complexity in nature. They also form the basis of most modern formulations of thermodynamics, both equilibrium and nonequilibrium. Today much progress is being made in our understanding of entropy and entropy generation in both fundamental aspects and application to concrete problems. The purpose of this book is to present some of these recent and important results in a manner that not only appeals to the entropy specialist but also makes them accessible to the nonspecialist looking for an overview of the field.

Although entropy has played an essential role in both equilibrium and nonequilibrium thermodynamics since the beginning, there are still many fundamental questions open. One of the most important is how to achieve an irreversible description of macroscopic phenomena when the underlying theory, quantum mechanics, is itself reversible. Kaufmann *et al.* address this question in their article *On Positivity of Rate of Entropy in Quantum-Thermodynamics* and find the answer in terms of quantum thermodynamics by using only restricted macroscopic information. Rateitschak *et al.* also use several concepts developed in information theory and nonlinear dynamics to study the *Entropy of Sequences Generated by Nonlinear Processes: the Logistic Map*. They find that higher-order entropies and the entropy per step are keys to the analysis of nonlinear processes and that quantities related to higher-order entropies can serve as measures of complexity. In his article *Developments in Normal and Gravitational Thermodynamics* Sewell reviews different facets of entropy governing the structure of equilibrium states in normal systems, where it is an extensive quantity, and in gravitational systems, where it is not. In the former case, he provides a precise specification of the conditions under which a set of macroscopic observables yield a complete thermodynamical description of a system. In the latter case, he discusses both non-relativistic systems, which support phases with negative *microcanonical* specific heat, and relativistic ones, involving black holes. In particular, he derives the generalised second law by an argument based exclusively on observable quantities and thus not involving the spurious notion of a black hole entropy.

Entropy generation is intimately related to problems of transport and energy conversion, particularly in the context of nonequilibrium thermodynamics. Indeed, transport phenomena were historically the first treated by nonequilibrium thermodynamics, and the derivation of the transport equations from the expression for the entropy generation illustrates the relation between entropy and transport. Since entropy generation implies dissipation, this relation immediately leads to the subject of the conversion of energy from one form to another. Of course energy can be converted between many other forms than those of heat and transport. Thus, it should come as no surprise that most of the articles of a more applied nature deal with transport and/or energy conversion. It is also understandable that many of the contributors to this volume could be described as thermodynamicists.

The relation between entropy generation and transport is particularly explicit in the form of nonequilibrium thermodynamics known as extended irreversible thermodynamics, although it is not restricted to transport problems. In *Extended Irreversible Thermodynamics: Statements and Prospects* Lebon *et al.* review the main statements of the theory, treat several examples, and discuss many of the questions which still remain to be resolved. J. Camacho (*Extended Thermodynamics and Taylor Dispersion*) uses extended irreversible thermodynamics to treat Taylor dispersion, a problem which is applicable to contaminant dispersion in rivers, for example. Woods treats *Higher Order Transport* and concludes that the constitutive equations can not be deduced from the condition that the entropy production be positive semidefinite. Muschik *et al.* develop the *Mesoscopic Theory of Liquid Crystals*.

Problems in energy conversion are especially amenable to treatment by finite-time thermodynamics. This form of nonequilibrium thermodynamics differs from most other formulations in that the irreversibilities are described macroscopically. Its concepts have been used to improve the performance of simulated annealing algorithms; the optimal path is the one producing least entropy. All this is discussed by Andresen in *Finite-time Thermodynamics and Simulated Annealing*. Biological systems may well be the ultimate energy conversion machines. In *Entropy, Information and Biological Materials* Cole argues that energy and order rejected by living entities at one level may be accepted by others at another lower level. He offers the hypothesis that information is conserved in the universe and that information lost during expansion after the "big bang" can be countered by a local increase in information by formation of more complex atomic/molecular arrangements which will show characteristics of elementary biological materials. Symmetry properties are of principal importance for energy conversion. Shiner and Sieniutycz demonstrate the validity of

PREFACE

a *Phenomenological Macroscopic Symmetry in Dissipative Nonlinear Systems* for a broad class of systems with arbitrary nonlinearities far from equilibrium.

Transport and energy conversion are often important aspects of problems of practical as well as fundamental importance. Cole discusses *The Provision of Global Energy* with emphasis on maximum possible work transfers within the restrictions of thermodynamics. The basic energy sources available – solar (including direct radiation, biomass, wind and waves), gravitational, geothermal and nuclear – and their suitability for global or for local applications are treated. This work provides a rational basis for the development of global energy policies. The conversion of solar energy to electricity in particular plays a prominent role in planning for the future provision of energy. In the direct conversion of radiant energy to electricity by semiconductor solar cells, it is found that the open-circuit voltage is less than the semiconductor energy gap by precisely the Carnot factor. In *Multiple Source Photovoltaics* Landsberg et al. show that this apparently fundamental feature is lost as soon as the usual single pump is replaced by several pumping sources. They also obtain other new results, including a novel form of the solar cell equation. The interrelations between transport, energy conversion and entropy generation become even more explicit in the article by Ratkje and Hafskjold, *Coupled Transport of Heat and Mass. Theory and Applications*, and Shiner et al. treat electrochemical systems in *Lagrangian and Network Formulations of Nonlinear Electrochemical Systems*.

The contributors to this volume were asked to either prepare a review of their recent work or present their most important new results in a manner which would make their contributions accessible not only to thermodynamicists and other specialists in their respective fields but also to other workers in the natural sciences. Each contribution was reviewed by at least two experts and the contributions revised accordingly. With this procedure it is hoped that a common failing of many multi-author books, namely that the level and quality of the individual contributions are very variable, has been avoided.

The contributions show that the volume promises to be of value to researchers both in academia and in industry, and to persons involved in the whole range of thermodynamics, from fundamentals to application. We hope that it will appeal to scientists from all disciplines - physicists, chemists, biologists - and engineers interested in fundamentals and applications of thermodynamics.

This book arose out of a workshop held in Spåtind, Norway in April 1994 and organized under the auspices of a research network of thermodynamicists supported by Human Capital and Mobility Contract No. CHRX-

Ct.92-0007 from the European Union. Some of the contributions were presented in preliminary form there, but others were solicited from outside the network. The concept of the book was further refined at the Gordon Research Conference on *Modern Developments in Thermodynamics*, held in Irsee, Germany in October of the same year. The contributors express their gratitude to the European Union as well as to the Gordon Research Conferences. Appreciation is also due to the Office of Naval Research and the International Science Foundation for their support of the Gordon Research Conference. Finally, as editor of this volume I am indebted to S.R. Berry, M. Davison, S. Sieniutycz, and L.C. Woods for their efforts in the realization of this work.

Bern, February 1996   J.S. Shiner

ON POSITIVITY OF RATE OF ENTROPY IN QUANTUM-THERMODYNAMICS

M. KAUFMANN, W. MUSCHIK AND D. SCHIRRMEISTER
Institut für Theoretische Physik
Technische Universität Berlin
Hardenbergstraße 36, 10623 Berlin, Germany

Abstract. The rate of entropy is discussed for several different dynamics of the generalized canonical operator: Canonical dynamics, Robertson dynamics, and contact time dynamics. For two discrete systems in contact the rate of entropy is positive definite, if the contact time is short, and if one of the two discrete systems is in equilibrium and the compound system of both is isolated.

1. Introduction

As it is well known quantum mechanics is a reversible theory [1]. Because quantum states of a system are described by its microscopic density operator ρ the fact of reversibility is expressed by an identical vanishing of the microscopical entropy rate

$$\dot{S}^0 \equiv 0, \qquad S^0 := -k\mathrm{Tr}(\rho \ln \rho) \tag{1}$$

(k is the Boltzmann constant).

An old question is how to obtain an irreversible description of processes. The answer is manifold. One possibility of irreversible description of systems is quantum-thermodynamics [2], a theory only using restricted macroscopic information of the considered system. This information about the system is achieved by a so-called restricted set of relevant (self-adjoint) observables

$$G_i \in \mathfrak{B}, \quad i = 1, 2, ..., n, \quad \mathfrak{G}_i = \mathfrak{G}_i^+, \tag{2}$$

$$\mathfrak{B} := (\mathfrak{G}_1, \mathfrak{G}_2, \mathfrak{G}_3, ..., \mathfrak{G}_n) \equiv (G). \tag{3}$$

1

J. S. Shiner (ed.), *Entropy and Entropy Generation*, 1–9.
© 1996 *Kluwer Academic Publishers*.

According to [3] we will denote this restricted set of relevant observables $\mathfrak{B}$ as *beobachtungsebene*. The choice of this beobachtungsebene is equivalent to the choice of a state space in case of a macroscopic thermodynamical system.

After having chosen the beobachtungsebene the microscopic density operator ρ of the system is not determined by the expectation values of the observables belonging to $\mathfrak{B}$

$$g :=< G >= \text{Tr}(\rho G) \tag{4}$$

because there are different microscopic density operators $\hat{\rho}$, $\text{Tr}\hat{\rho} = 1$, satisfying also the n relations (4)

$$g = \text{Tr}(\hat{\rho}G) = \text{Tr}(\rho G). \tag{5}$$

Consequently the different density operators $\hat{\rho}$ and ρ are equivalent with respect to $\mathfrak{B}$. Because ρ is not determined by the set of the g we need an additional principle for choosing the density operator by which the system should be described in accordance with the restricted knowledge we have according to (4). The well-known procedure stems from Jaynes [4, 5]: We choose that density operator R which maximalizes the (macroscopic) entropy of the system

$$S := -k \min_{\hat{\rho}} \text{Tr}(\hat{\rho}\ln \hat{\rho}) = -k\text{Tr}(R\ln R), \tag{6}$$

$$g = \text{Tr}(RG) = \text{Tr}(\rho G), \qquad \text{Tr}R = 1. \tag{7}$$

The maximalization (6) taking into account the constraints (5) yields the well-known result that R has the form of the generalized canonical operator [6]

$$R = Z^{-1}\exp(-\boldsymbol{\lambda} \cdot \boldsymbol{G}), \qquad Z := \text{Tr}\exp(-\boldsymbol{\lambda} \cdot \boldsymbol{G}). \tag{8}$$

Here the n quantities $\boldsymbol{\lambda}$ have to be determined by the n constraints (5).

2. Dynamics

Time derivation of (6) yields for the entropy rate of the total system [7]

$$\dot{S} = -k\text{Tr}(\dot{R}\ln R). \tag{9}$$

Consequently to determine the rate of entropy we need the dynamics of the considered system represented by $\dot{R}$. There are now different possibilities of introducing $\dot{R}$.

2.1. CANONICAL DYNAMICS

The first possibility presupposes that R has canonical form (8) for all times. Therefore (9) yields by use of (8)

$$\dot{S} = k\mathrm{Tr}(\lambda \cdot G\dot{R}). \tag{10}$$

$\dot{R}$ is determined by the time rates of the λ appearing in (8) and by the time rates of the work variables a on which the observables G in $\mathfrak{B}$ depend. Using the first equation in (7) the $\dot{\lambda}$ can be replaced by the $\dot{g}$. In equilibrium the rates $\dot{a}$ and $\dot{\lambda}$, or $\dot{g}$ respectively, vanish, but $\dot{R}$ is not zero due to quantum-mechanical dynamics: In equilibrium R satisfies the v.Neumann equation. Thus we have

$$\dot{R} = -\imath LR + (\partial R/\partial a) \cdot \dot{a} + (\partial R/\partial \lambda) \cdot \dot{\lambda}, \tag{11}$$

This type of dynamics is called *canonical*. The coefficients $\partial R/\partial a$ and $\partial R/\partial \lambda$ can be calculated by differentiation of (8) [7]. The expressions

$$v := -\imath \mathrm{Tr}(GLR) \tag{12}$$

are called *quantum-mechanical drift terms*. They originate from the first part of $\dot{R}$ in (11).

For demonstration we discuss a simple example. We consider the Hamiltonian of a closed discrete system S which is separated from its vicinity V by an partition ∂S

$$\mathcal{H} = \mathcal{H}^S(a) + \mathcal{H}^V(a) + \mathcal{H}^{\partial S}(a) \tag{13}$$

$$[\mathcal{H}^S, \mathcal{H}^V] = 0, \qquad \dot{\mathcal{H}} = 0. \tag{14}$$

Here $\mathcal{H}^S$ is the Hamiltonian of the discrete system in consideration, $\mathcal{H}^V$ the Hamiltonian of its vicinity, and $\mathcal{H}^{\partial S}$ that of the interaction between the system and its vicinity. Additionally we presuppose that ∂S is inert [7]

$$\mathrm{Tr}(\dot{\mathcal{H}}^{\partial S} R) = 0, \qquad \mathrm{Tr}(\mathcal{H}^{\partial S} \dot{R}) = 0. \tag{15}$$

We now choose the beobachtungsebene of that discrete compound system

$$\mathfrak{B}^0 = (\mathcal{H}, \mathcal{H}^S, \mathcal{H}^V) \tag{16}$$

to which, according to (3) and (8), the following generalized canonical operator belongs

$$R = (1/Z) \exp\{-(\lambda^{\mathcal{H}}\mathcal{H} + \lambda^{\mathcal{H}^S}\mathcal{H}^S + \lambda^{\mathcal{H}^V}\mathcal{H}^V)\}. \tag{17}$$

According to (10) the rate of entropy is by use of canonical dynamics

$$\dot{S} = k\mathrm{Tr}\{(\lambda^{\mathcal{H}}\mathcal{H} + \lambda^{\mathcal{H}^S}\mathcal{H}^S + \lambda^{\mathcal{H}^V}\mathcal{H}^V)\dot{R}\}. \tag{18}$$

Presupposing an inert partition (15) this expression can be transformed by introducing the heat exchanges with respect to S and V [7]

$$\dot{Q}^S := \mathrm{Tr}(\mathcal{H}^S\dot{R}), \quad \dot{Q}^V := \mathrm{Tr}(\mathcal{H}^V\dot{R}) = -\dot{Q}^S \tag{19}$$

into

$$\frac{1}{k}\dot{S} = (\lambda^{\mathcal{H}^S} - \lambda^{\mathcal{H}^V})\dot{Q}^S. \tag{20}$$

Thus for this simple non-equilibrium example the rate of entropy is proportional to the heat exchange which is here the only source of entropy. The interpretation of the factor $\lambda^{\mathcal{H}^S} - \lambda^{\mathcal{H}^V}$ in (20) is postponed to section 3.

In the first part of this section $\dot{R}$ was not inserted into (11), but the heat exchanges (19) were introduced. We now use (11) to calculate (10). We obtain [8]

$$\dot{S}_{can} = -k\lambda \cdot \mathcal{K} \cdot \dot{\lambda} - k\lambda \cdot (\Delta\dot{G} \mid \Delta G) \cdot \lambda. \tag{21}$$

Here the following abbreviations were introduced: $\mathcal{K}$ is the symmetric, positive definite canonical correlation matrix

$$\mathcal{K} := (\Delta G \mid \Delta G), \tag{22}$$

and $(F \mid M)$ the generalized Mori product [9]

$$(F \mid M) := \int_0^1 du\, \mathrm{Tr}(RF^+ R^u M R^{-u}) \tag{23}$$

The symbol Δ means the deviation operator defined by

$$\Delta X := X - \mathrm{Tr}(RX). \tag{24}$$

The dependence of the entropy rate in canonical dynamics of the $\dot{\lambda}$ and of the $\dot{a}$ is given by (21) for an unspecified beobachtungsebene of observables which depend on work variables.

2.2. GENERALIZED ROBERTSON DYNAMICS

A different approach can be achieved by using the fact that R can be represented by a projection of ρ [10]

$$R(t) = P(t)[\rho(t)], \quad P^2(t) = P(t). \tag{25}$$

Time differentiation yields

$$\dot{R}(t) = \{P(t)[\rho(t)]\}^{\bullet} = \dot{P}(t)[\rho(t)] + P(t)[\dot{\rho}(t)]\}, \tag{26}$$

and dynamics is determined by $\dot{P}$. If the projection operator P is especially chosen as the Kawasaki-Gunton operator [11]

$$K(t)[*] := R(t)\mathrm{Tr}(*) + (\partial R/\partial g) \cdot (\mathrm{Tr}(G*) - g\mathrm{Tr}(*)) \tag{27}$$

the representation (25) of R is identical for all times with its canonical form (8) according to (7). Consequently canonical dynamics and so-called (generalized) *Robertson dynamics* [12] defined by $\dot{P}(t)[*] \equiv \dot{K}(t)[*]$ are identical. Calculating (26) we obtain, if we write

$$K(t)[AB] + \dot{K}(t)[CB] \equiv (KA + \dot{K}C)B(t), \tag{28}$$

the following expressions [13]

$$\dot{R}(t) = -\imath(KL + \imath\dot{K})(t)R(t) - \int_{t_0}^{t} ds H(s, t)R(s), \tag{29}$$

$$H(t, s) := (KL + \imath\dot{K})(t)T(t, s)(QL - \imath\dot{K})(s), \tag{30}$$

$$Q \equiv Q(t)[*] := 1 - K(t)[*], \tag{31}$$

$$\frac{\partial}{\partial s}T(t, s) = \imath T(t, s)(QL - \imath\dot{K})(s), \quad T(t, t) = 1. \tag{32}$$

Here the initial preparation

$$\rho(t_0) = R(t_0) \tag{33}$$

was used which is essential to get rid of the initial conditions in (29).

Calculating the rate of entropy in Robertson dynamics according to (9) and (29) we obtain [8]

$$\dot{S}_{Rob} = k \int_{t_0}^{t} ds\mathrm{Tr}[(\lambda \cdot G)(t)(KL + \imath\dot{K})(t)T(t, s)(QL - \imath\dot{K})(s)R(s)]. \tag{34}$$

2.3. CONTACT TIME DYNAMICS

If the considered system is in contact with a vicinity, and if the contact time between them is sufficiently short (so that conduction problems are out of scope), Robertson dynamics transforms into the special case of *contact time dynamics*.

If we use the special beobachtungsebene (16) of the considered compound system, we obtain [8] from generalized Robertson dynamics (34) an approximation for small contact times $\Delta t := t - t_0$ between the parts of the compound system

$$\dot{R} = \dot{K}R - (KL + \imath\dot{K})(L - \imath\dot{K})R\Delta t \tag{35}$$

In the last section we discuss this dynamics especially with respect to the positive definiteness of the rate of entropy.

3. Positive rate of entropy

First of all we use a phenomenological argument to illustrate that the rate of entropy is positive definite. For that purpose we remember: Temperature is only defined for equilibrium states and therefore only for reversible processes [14]. Thus a redefinition of temperature with regard to nonequilibrium processes is necessary. This dynamical analogue of the thermostatic temperature is the so-called *contact temperature* [15, 16]. The defining inequality of the contact temperature Θ^S of a closed discrete system S whose partition ∂S is impervious to power exchange runs as follows

$$\lim_{t \to \epsilon} \left(\frac{1}{\Theta^S(t)} - \frac{1}{T^V(0)} \right) \dot{Q}^S(t) \geq 0, \quad 2\eta > \epsilon > \eta \to 0. \tag{36}$$

Here $T^V(0)$ is the thermostatic temperature of the system's vicinity which is supposed to be in equilibrium at time $t = 0$, when the system gets in contact with its vicinity. The interpretation of the contact temperature is very easy: The defining inequality (36) determines the contact temperature $\Theta^S(t)$ as that thermostatic temperature of the system's vicinity so that the net heat exchange between them vanishes.

Thus we get

$$\Theta^S(t) \geq T^V(t) \quad \longrightarrow \quad \dot{Q}^S(t) \leq 0, \tag{37}$$
$$\Theta^S(t) \leq T^V(t) \quad \longrightarrow \quad \dot{Q}^S(t) \geq 0 \tag{38}$$

which is the statement of (36).

An interpretation of (20) with regard to the Second Law is now possible [17, 18] : If we identify

$$\lambda^{\mathcal{H}^S}(t) + \Lambda(t) \equiv \frac{1}{k\Theta^S(t)}, \quad \lambda^{\mathcal{H}^V}(t) + \Lambda(t) \equiv \frac{1}{kT^V(t)}, \tag{39}$$

then from (20) and (36)

$$\dot{S} \geq 0 \tag{40}$$

follows by comparison. Therefore S represents a nonequilibrium system which is separated by an inert, diathermal, and power-isolating partition

∂S from its surrounding vicinity V, the latter being always in equilibrium (reservoir). The meaning of $\Lambda(t)$ is explained elsewhere [7] and is here of no interest. The result we obtain is the following: The entropy rate (20) is valid for a special, but sufficiently general beobachtungsebene (16) by use of canonical dynamics it is positive definite, if the heat exchange between parts of the compound system is interpreted by introducing the contact temperature which is defined by the inequality (36). But the interesting question is just the other way round: What are the presuppositions to prove the positive definiteness of the entropy rate without use of contact temperature and what is then the quantum-thermodynamical interpretation of the contact temperature ?

Up to now there is no way to prove the positivity of $\dot{S}_{can}$ in (21) or of $\dot{S}_{Rob}$ in (34). The conjecture is that these quantities are not definite unless additional presuppositions are made. One possibility to enforce positivity is to make a relaxation ansatz for the λ in case of constant work variables

$$\dot{G} \equiv 0, \qquad \dot{\lambda} = -A(\lambda) \cdot \lambda, \quad A \text{ positive semi-definite.} \qquad (41)$$

Then (21) yields

$$\dot{S}_{can} = \lambda \cdot \mathcal{K} \cdot \mathcal{A} \cdot \lambda \geq 0 \qquad (42)$$

because of the positivity of $\mathcal{K}$ and A. But first of all there is no reason to demand (41).

In case of contact time dynamics (35) we can prove the following

□*Proposition* [8]: If all quantum-mechanical drift terms (12) vanish for the chosen beobachtungsebene, then we obtain by use of contact time dynamics for the entropy rate of the isolated compound system

$$\dot{S}_{ct} = k(\imath\lambda \cdot LG + \lambda \cdot \tilde{Q}\dot{G} \mid \imath\lambda \cdot LG + \lambda \cdot \tilde{Q}\dot{G})\Delta t \geq 0, \qquad (43)$$

$$\tilde{Q} := 1 - \tilde{P}, \qquad \tilde{P} := \mid \Delta G) \cdot \mathcal{K}^{-1} \cdot (\Delta G \mid . \qquad (44)$$

□

For deriving the positivity of the entropy production (40) the validity of the defining inequality (36) was presupposed. The proposition (43) now allows to derive an inequality which can be interpreted as the defining inequality (36). In case of constant work variables the rate of entropy in contact time dynamics (43) becomes

$$\dot{a} \equiv 0, \rightarrow \dot{S}_{ct} = k(\imath\lambda \cdot LG \mid \imath\lambda \cdot LG)\Delta t \geq 0. \qquad (45)$$

We now consider the discrete compound system described by the Hamiltonian (13) and by the beobachtungsebene (16). The partition between both the sub-systems is presupposed as being inert (15). Then (45) becomes

$$\dot{S}_{ct} = k(\lambda^{\mathcal{H}}\imath L\mathcal{H} + \lambda^{\mathcal{H}^S}\imath L\mathcal{H}^S + \lambda^{\mathcal{H}^V}\imath L\mathcal{H}^V \mid \imath\lambda \cdot LG)\Delta t. \qquad (46)$$

The heat exchanges (19) are in contact time dynamics

$$\dot{Q}^S = (\imath L\mathcal{H}^S \mid \imath\lambda \cdot LG)\Delta t, \quad \dot{Q}^V = (\imath L\mathcal{H}^V \mid \imath\lambda \cdot LG)\Delta t. \tag{47}$$

Because the first term in (46) is zero we rediscover (20)

$$\dot{S}_{ct} = k(\lambda^{\mathcal{H}^S} - \lambda^{\mathcal{H}^V})\dot{Q}^S \geq 0. \tag{48}$$

This inequality is now compared with the defining inequality of the contact temperature (36). After a short reflection we obtain the result

$$1/\Theta^S = \alpha\lambda^{\mathcal{H}^S} + c, \quad 1/T^V = \alpha\lambda^{\mathcal{H}^V} + c, \tag{49}$$

$$\alpha = konst > 0, \qquad c = const. \tag{50}$$

A comparison with (39) demonstrates that

$$\alpha = k, \quad \text{and} \quad c = k\Lambda \tag{51}$$

are valid. Therefore the connection between contact time dynamics and the defining inequality of contact temperature is evident.

4. Discussion

The choice of a state space for a thermodynamical discrete system is replaced in quantum-thermodynamics by introducing the beobachtungsebene, that is the restricted set of relevant observables. By this choice of the beobachtungsebene and by maximalization of the system's entropy with respect to the constraints the microscopic density operator is substituted by the (generalized) canonical (density) operator which describes all properties of the system with respect to the beobachtungsebene exactly also in non-equilibrium. Because the construction of the canonical operator is here local in time (belongs e.g. to the initial time) the question arises what dynamics the canonical operator satisfies.

Two different views are possible: Because firstly the canonical operator includes the work variables and the Lagrange parameters due to the maximalization its dynamics is determined by the time rates of the work variables and of the Lagrange parameters. This kind of dynamics which preserves the form of the canonical operator for all times is called canonical. Because secondly the canonical operator can be represented as a projection of the microscopic density operator its dynamics is determined by the time rate of the projection operator and by the system's Hamiltonian. This dynamics is denoted as Robertson dynamics. If now the projection is especially performed by the Kawasaki-Gunton operator, the form of the canonical operator is always preserved, so that in this case canonical and Robertson dynamics are equivalent.

Starting out with the exact expressions for the rate of entropy in canonical and in Robertson dynamics we need restricting assumptions to prove its positivity: In canonical dynamics in case of constant work variables we need a relaxation ansatz for the Lagrange parameters. But there is no reason that the Lagrange parameters obey such a relaxation ansatz. In Robertson dynamics the contact time approximation results in the positivity of the entropy rate also in the case of time dependent work variables and unspecified beobachtungsebene of vanishing quantum-mechanical drift terms. For the case of a non-equilibrium system in contact with an equilibrium vicinity the Lagrange parameters are interpretable: They are proportional to the contact temperature of the non-equilibrium system and to the thermostatic temperature of the equilibrium vicinity. Thus a quantum-thermodynamical foundation of the contact temperature is possible, and the positivity of the rate of entropy can be proved for contact time dynamics.

References

1. Dougherty, J.P. (1994) *Phil. Trans. R. Soc. Lond.* A **346**, 259–305.
2. Muschik, W. and Kaufmann, M. (1993) in W. Ebeling and W. Muschik (eds.) *Statistical Physics and Thermodynamics of Nonlinear Nonequilibrium Systems - Statistical Physics 18 Satellite Meeting*, World Scientific, Singapore, pp. 229–242.
3. Schwegler, H. (1955) *Z. Naturforsch* **20a**, 1543–1553.
4. Jaynes, E.T. (1957) *Phys. Rev.* **106**, 620.
5. Jaynes, E.T. (1957) *Phys. Rev.* **108**, 171.
6. Jaynes, E.T. (1979) in R. D. Levine and M. Tribus (eds.) *The Maximum Entropy Formalism*, MIT Press, Cambridge.
7. Muschik, W. and Kaufmann, M. (1994) *J.Non-Equilib.Thermodyn.* **19** 76–94.
8. Schirrmeister, D. (1994) Diplomarbeit, Institut für Theoretische Physik, Technische Universtität Berlin.
9. Mori, H. (1965) *Progr. Theor. Phys.* **34**, 399.
10. Fick, E. and Sauermann, G. (1983) *Quantenstatistik dynamischer Prozesse (vol. 1)*, Verlag Harri Deutsch, Thun, Frankfurt/Main.
11. Kawasaki, K. and Gunton, J. D. (1973) *Phys. Rev.* **A8**, 2048.
12. Robertson, B. (1966) *Phys. Rev.* **144**, 151.
13. Balian, R., Achassid, Y. and Reinhardt, H. (1986) *Physics Reports* **131**, 1–146.
14. Muschik, W. (1977) *Arch. Rat. Mech. Anal.* **66**, 379–401.
15. Muschik, W. and Brunk, G. (1977) *Int. J. Engng. Sci.* **15**, 377.
16. Muschik, W. (1981) in O. Brulin and R.K.T. Hsieh (eds.) *Continuum Models of Discrete Systems 4*, North-Holland, Amsterdam, p. 511.
17. Muschik, W. (1990) *Aspects of Non-Equilibrium Thermodynamics, 6 Lectures on Fundamentals and Methods*, World Scientific, Singapore, Sect. 4.1.
18. Muschik, W. and Fang, J. (1989) *Acta Phys. Hung.* **66**, 39–57.

ENTROPY OF SEQUENCES GENERATED BY NONLINEAR PROCESSES: THE LOGISTIC MAP

K. RATEITSCHAK, J. FREUND AND W. EBELING

Institute of Physics, Humboldt–University Berlin
Invalidenstraße 110, D-10099 Berlin, Germany

1. Introduction

Nonlinear processes play an important role in many sciences like physics, chemistry, biology, medicine, economy, ecology etc. Very often data series are given and one is interested in connecting them to a class of processes. The modelling of such by low dimensional dynamical systems is favourable and physically meaningful.

The reduction of a high–dimensional system to an effectively low–dimensional one is possible through coherent collective behaviour. Dissipative nonlinear systems which are driven by some external pumping mechanism often posses an attractor which is different from a point but nevertheless of comparatively low dimension. Macroscopic dynamical modes, e.g. Bénard convection cells, emerge when the dissipation rate, here related to the temperature gradient, exceeds a critical threshold value. As a result the dynamics can be effectively described by only a few degrees of freedom.

The method of Poincaré section [1] allows a further reduction of the dimensionality of the effective motion and additionally marks the way to discrete dynamical systems. Thus one is lead to one–dimensional maps which still are intimately related to dissipative physical systems [2]. One–dimensional discrete maps have served as key systems for the understanding of many by now standard features like e.g. self-similarity, the period-doubling cascade, intermittency, etc., which are found in the realm of nonlinear dynamics, chaos and complex systems [3]. In the present paper we regard the prominent logistic map as a token of a microscopically formulated effective law of motion which is time discrete - possibly as a consequence of stroboscopic observation of a time continuous system - and deterministic.

11

J. S. Shiner (ed.), Entropy and Entropy Generation, 11–26.
© *1996 Kluwer Academic Publishers.*

A partitioning of state space into a finite number of boxes labelled with symbols means a coarse–grained description of the process. As a consequence the time discrete microscopic state space trajectories are mapped onto symbol sequences related to a macroscopic level of description. This procedure generally is called symbolic dynamics [4].

The statistical characteristics of the stationary state, established already at the microscopic level by the notion of the Gibbsian ensemble, are then investigated at the macroscopic level. The statistical analysis of a sample string is equivalent to projecting the microscopic statistics into the coarse-grained space if the system is ergodic and mixing [5]. Peculiarities of the microscopic dynamics are reflected in correlations existing between the symbols. These correlations give rise to characteristic statistics, i.e. probability distribution $p(A_n)$ of subsequences or n-words $A_n := a_1, ..., a_n$; here, all a_i are letters of the alphabet $\{c_1, ..., c_\lambda\}$. If the partition of the microscopic space is appropriately chosen, namely as a generating partition [6], it is even possible to extract valuable information about the microscopic system by analyzing the statistics of macrostates.

The measures we employ for the statistical analysis of the symbol strings are dynamical entropies. The entropy concept is closely related to some prominent quantities, e. g. Lyapunov exponent or Kolmogorov–Sinai entropy [6], designed for a description of chaotic systems. One can classify nonlinear dynamical systems by considering their entropy profile. Some processes of the aforementioned sciences mark the border between order and chaos [7, 8, 9].

At this point we define the dynamical entropies considered in this work. The n–block entropy

$$H_n := -\sum_{A_n} p(A_n) \log p(A_n) . \tag{1}$$

is an informational measure, which is applied to the probability distribution $p(A_n)$. Although the calculation of these numbers from the probability distribution means a reduction characteristic features of the dynamical system are still reflected by the series of H_n (for $n = 1, 2, 3, ...$). The average uncertainty per symbol is $H(n) := H_n/n$. Another useful quantity is the conditional entropy defined by:

$$h_n = H_{n+1} - H_n . \tag{2}$$

The decay rates of the h_n allow conclusions about the strength and effective range of correlations between the symbols and can be used to define a memory [10].

McMillan and Khinchin have shown that for the stationary and ergodic source the limit

$$h := \lim_{n \to \infty} H(n) = \lim_{n \to \infty} h_n . \tag{3}$$

exists; h is named entropy of the source [11, 12].

In connection with the McMillan theorem [12] has defined an effective number of words $N^*(n)$ [11]

$$N^*(n) = \lambda^{hn}. \tag{4}$$

Of special interest for our investigations will be processes showing a long slowly decaying memory. That means the decay rate of the conditional entropies follows a power law: $h_n \sim 1/n^\alpha$. In passing by we note that generically one finds an exponential law [10] which indicates rapidly decaying correlations, hence a short memory.

Besides the method of symbolic dynamics natural symbol sequences like DNA, text and music are also found as the result of evolutionary processes. Thus it is intriguing to inquire about a relationship between natural symbol sequences and nonlinear dynamical systems. It has been conjectured by Ebeling and Nicolis that natural symbol sequences result from nonlinear dynamical systems located at the borderline between order and chaos [8, 9].

In the past two decades there have been found three prominent routes to chaos: intermittency, via quasiperiodicity and the period doubling scenario [13].

Szépfalusy and Győrgi have shown that symbol sequences related to a special weakly intermittent system leads to $h_n \sim 1/n^\alpha, \alpha > 2.5$ with $h > 0$ [14].

Self similar binary sequences, for instance, result from the limit of the period doubling scenario (logistic map) and from the limit of the quasiperiodicity route (critical circle map). They have been studied by Grassberger [15], Gramss [16], and ourselves [9, 17]. The conditional entropies show a power law decay $h_n \sim 1/n^\alpha, \alpha = 1$ with $h = 0$.

Numerical analyses of text and music sequences confirmed the conjecture of Ebeling and Nicolis. It was found that $h_n \sim 1/n^\alpha$ with $0 < h \ll 1$ and $\alpha = 0.5$ for texts and $0.5 \leq \alpha \leq 1$ for music [9, 18, 19, 20]. Exact exponents are difficult to extract from the empirical material. A nonlinear dynamical model for texts with a scaling law $h_n \sim 1/n^\alpha \ \alpha < 1$ and $h = 0$ is introduced in [21].

With respect to their statistical properties DNA sequences are quite different from text and music. Informational analyses of biosequences can be found in [22, 23].

Applying Grassberger's effective measure complexity (EMC) defined by [15]:

$$EMC := \sum_{n=0}^{\infty} (h_n - h) \tag{5}$$

to self similar sequences, text and coded music yields the value infinity.

In this paper we study the symbolic dynamics of the period doubling accumulation point (also called Feigenbaum accumulation point) of the logistic map in more detail. First results concerning its informational analysis we already mentioned above. In the second section we introduce a symbol sequence generator as an equivalent approach to the resulting binary symbol sequence involving grammatical rules in the context of formal languages [24]. In the third section we analytically derive a formula for the block entropies based on the symbol sequence generator. In the fourth section we extend our considerations by choosing arbitrary partitions of the state space.

2. Symbolic Dynamics of the Feigenbaum Accumulation Point

The perhaps most prominent nonlinear map is the logistic map, defined by

$$x_{n+1} = f(x_n) = rx_n(1 - x_n) \qquad r\epsilon[0,4] \,. \tag{6}$$

The logistic map was introduced by Verhulst already in 1845 as a model for the growth of a population in an closed area. The logistic map shows a rich dynamic behaviour, like period doubling, intermittency and crisis, for values of the control parameter r varying in the range between 1 and 4 [13]. Starting with $r = 1$ and increasing r one can observe a route from order to chaos: the period doubling scenario [25]. The transition point between multiperiodicity and chaos is called the Feigenbaum accumulation point $r_\infty = 3.56994567\ldots$ and will be the object of our entropy analysis. The microscopic probability density of the Feigenbaum accumulation point has a Cantor structure [25]. That means the attractor is selfsimilar.

Choosing the generating partition: $[0, 0.5] \to 0$ and $(0.5, 1] \to 1$ the selfsimilar attractor leads to a selfsimilar binary symbol sequence.

First investigations concerning the statistics of these binary sequences were done by Grassberger [15]. He reported a formula for the block entropies for $n = 2^k$ (for $k = 1, 2, 3, \ldots$)

$$H_{n=2^k} = \log_2\left(\frac{3n}{2}\right) \tag{7}$$

Ebeling and Nicolis [9] completed this formula for all word lengths n. The basis of their results were empirical rules derived by observing typical sequences. The fundamental result is an overall decay of the h_n according to a

$$h_n \sim \frac{1}{n} \qquad \text{law with } h = 0 \,. \tag{8}$$

In agreement with the McMillan theorem [11, 12] all occurring words are standard words and the effective number of word (4) resp. the number of standard words grows slower than exponentially [26], in this case namely $N^*(n) \sim n$ [9].

The rank ordered word distributions (r. o. w. d.) give a first hint at a selfsimilar sequence [17]. For word lengths $n = 2^k$ the r. o. w. d. is a box of length $\frac{3}{2}n$. For word lengths $2^k < n < 2^{k+1}$ the r. o. w. d. consists of two boxes.

The replacement rules

$$1 \rightarrow 10$$
$$0 \rightarrow 11 \tag{9}$$

applied to an infinitely long sequence yields the same sequence. This rescaling invariance is the defining property of selfsimilarity. These replacement rules establish a link to grammatical rules in the context of formal languages [24]. We constructed the following deterministic symbol sequence generator producing the binary Feigenbaumsequence on basis of the replacement rules (9) [27]:

$$\begin{aligned} a_0 &= 1 \\ a_1 &= 10 \\ a_{n+1} &= a_n \circ a_{n-1} \circ a_{n-1}. \end{aligned} \tag{10}$$

Here a_i denotes a symbol sequence of length 2^i. The symbol $\circ$ means concatenation of two subsequences.

Another example of such a system is a generator producing the selfsimilar "rabbit sequence" [16, 28]. There are also systems involving stochastic rules beside deterministic ones to create symbol sequences with long range correlations [21, 27, 29, 30].

Another way to produce the Feigenbaumsequence is the following: We start with the symbol 0 and use the replacement rule $0 \rightarrow 11$. We append the sequence 11 behind the symbol 0. Now we read the second symbol 1 and use the replacement rule $1 \rightarrow 10$. We append the sequence 10 behind the last 1 of our sequence. Now we read the third symbol 1 and so on.

By continuation this leads to the Feigenbaumsequence produced by the symbol sequence generator except for the first symbol 1. This construction method allows the generation of the Feigenbaumsequence through a set of binary switches (figure 1).

The construction of the Feigenbaumsequence through a set of binary switches can be linked to coding theory. The main task in the theory of coding [11, 31] is to find the shortest representation of a given massage (symbol sequence). The algorithmic complexity K corresponds to the length of the shortest program $\mathcal{P}(T)$ capable of producing the message of length

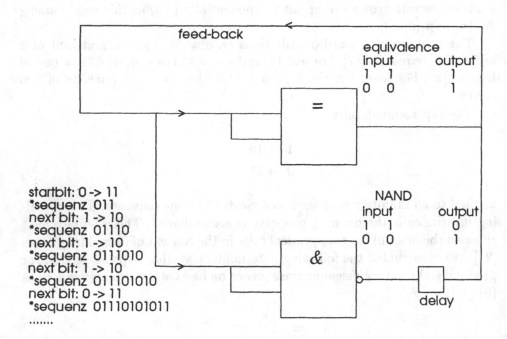

Figure 1. Generation of the Feigenbaumsequence through a set of binary switches

T on an universal Turing machine [32, 33]:

$$K = \min_{\mathcal{P}} |\mathcal{P}(T)| \,. \qquad (11)$$

Here, $|.|$ denotes the binary length of the program.

The Kolmogorov–Sinai entropy and the algorithmic complexity are related to each other by [4]

$$KS = \lim_{T \to \infty} \frac{K(T)}{T} \,. \qquad (12)$$

The binary length of the program to construct the binary switches and the start symbol 0 is finite. That means the algorithmic complexity K is finite. (12) yields $KS = 0$ in correspondence with former results (for a generating partition $KS = h$ [6], Feigenbaum accumulation point $h = 0$ [9]). Another possibility to produce the Feigenbaumsequence is the symbol sequence generator (10). The binary length of data necessary to code this generator and the start sequences a_0 and a_1 is finite. That means the algorithmic complexity K is finite in accordance with the binary switch representation (figure 1) but in contrast to the measure theoretical complexity EMC which diverges.

The application of algorithmic complexity K and of measure theoretical complexity EMC to the Feigenbaumsequence reveals once more that complexity is a question of definition. This already was pointed out by Grassberger [15, 34]. For a discussion of diverse complexity measures see [35, 36].

3. Derivation of the Entropy Formula

In this section we analytically derive a formula for the n–word probability distributions and for the block entropies for all word lengths n on basis of the symbol sequence generator (10).

Our generator has the following properties:

Lemma 1: The sequences a_z and $a_{z-1}a_{z-1}$ only differ in the last symbol.
Proof:

$$
\begin{aligned}
a_z &= a_{z-1}a_{z-2}a_{z-2} \\
&= a_{z-1}a_{z-2}a_{z-3}a_{z-4}a_{z-4}
\end{aligned}
$$

without general restriction z-even

$$
\begin{aligned}
&= \prod_{i=1}^{z-1} a_{z-i}a_0 a_0 \\
&= \prod_{i=1}^{z-1} a_{z-i}11 \qquad (13)
\end{aligned}
$$

$$
\begin{aligned}
a_{z-1}a_{z-1} &= a_{z-1}a_{z-2}a_{z-3}a_{z-3} \\
&= a_{z-1}a_{z-2}a_{z-3}a_{z-4}a_{z-5}a_{z-5} \\
&= \prod_{i=1}^{z-1} a_{z-i}a_1 \\
&= \prod_{i=1}^{z-1} a_{z-i}10 \qquad (14)
\end{aligned}
$$

q.e.d.

Lemma 2: Discarding the first 2^{z-x} symbols the new resulting sequence is the same up to the string position $2^{z-x} - 1$.
Proof:

$$
\begin{aligned}
a_{z+1} &= a_z a_{z-1}a_{z-1} \\
&= a_{z-x} \prod_{i=x+1}^{2} (a_{z-i}a_{z-i})a_{z-1}a_{z-1} \\
&= a_{z-x}a_{z-x-1}a_{z-x-1} \prod_{i=x+2}^{2} (a_{z-i}a_{z-i})a_{z-1}a_{z-1} \qquad (15)
\end{aligned}
$$

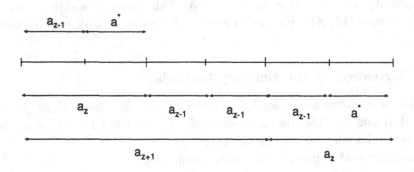

a_{z-1} and $a^{\bullet}$ $=a_{z-2}\,a_{z-2}$ differ only in the last bit

Figure 2.   sequence positions

Now a_{z-x} will be discarded. From Lemma 1 it follows that the new sequence is the same up to the string position $2^{z-x} - 1$.

q.e.d.

Lemma 3: For word lengths $n \leq 2^{z-1}$ the words which start at the sequence position $2^{z-1} + i$ and $2^{z-1} + 2^{z-2} + i$ for all $i = 1 \ldots 2^{z-2}$ are identical.

Proof:

$$a_z a_{z-1} = a_{z-1} a_{z-2} a_{z-2} a_{z-1} \tag{16}$$

$a_{z-2} a_{z-2}$ and a_{z-1} differ only in the last bit (Lemma 1). The last bit of a_{z-1} will be not reached.

q.e.d.

As a consequence of Lemma 2 the first $2^{z-1} - 1$ 0 and 1 repeat after position 2^{z-1} and so it is possible to derive an iteration formula for the frequency $\sigma(n, 2^{z+1})$ of a word of length $2^{z-2} < n \leq 2^{z-1}$ in a sequence of the length 2^{z+1} (figure 2):

The first bit of the word of length n may arbitrarily occur in a_{z+1}. Counting the word in a_z and twice in a_{z-1} one gets the same probability

$$\sigma(n, 2^{z+1}) = \sigma(n, 2^z) + 2\sigma(n, 2^{z-1}) . \tag{17}$$

Now we determine the initial conditions (figure 3):

Let be $n = 2^{z-2}$. The first $2^{z-2} + 2^{z-3}$ words are different, but up to now we regard this as an initial assumption. The last 2^{z-3} words repeat after position $2^{z-2} + 2^{z-3}$ (Lemma 3).

Now let be $n = 2^{z-2} + i$ with $i = 1, \ldots, 2^{z-2}$. As a consequence of Lemma 2 the first $2^{z-1} - 1$ symbols repeat after position 2^{z-1}. It follows

```
      n=4                      n=5                          n=6

1011101010111011        101110101011101110111        101110101011011101110
1011    new             10111   new                  101110  new
0111    new             01110   new                  011101  new
1110    new             11101   new                  111010  new
1101    new             11010   new                  110101  new
-----------------       ----------------------       -----------------------
1010    new             10101   new                  101010  new
0101    new             01010   new                  010101  new
1010    old             10101   old                  101011  new
0101    old             01011   new                  010111  new
-----------------       ----------------------       -----------------------
1011    old             10111   old                  101110  old
                        01110   old                  011101  old
                        11101   old                  111011  new
                        11011   new                  110111  new
                        ----------------------       -----------------------
                        10111   old                  101110  old
```

Figure 3.   occurrence of the words

that the first $2^{z-1} - n$ words repeat after position 2^{z-1}. The other words starting between the positions 2^{z-1} and $2^{z-1}+2^{z-2}+1$ are new, because the symbols at positions 2^{z-1} and 2^z are different. As a consequence of Lemma 3 the words starting between positions $2^{z-1}+2^{z-2}$ and 2^z+1 are the same as the words starting between positions 2^{z-1} and $2^{z-1}+2^{z-2}+1$. Of the words which occur for $n = 2^{z-2}$ twice between 2^{z-2} and $2^{z-1}+1$, only $2^{z-2}+2^{z-3} - n$ words occur twice. The other words reach the last position of a_{z-1} (Lemma 3). If $n > 2^{z-2}+2^{z-3}$ no word occurs twice between 2^{z-2} and $2^{z-1}+1$. It follows that the first $2^{z-1}+2^{z-2}$ words are different for $n = 2^{z-1}$ (initial assumption).

We distinguish between two groups:

Group X: $2^{z-2}+2^{z-3} \leq n \leq 2^{z-1}$ initial conditions:

$$\sigma_a(n,2^z) = 2 \qquad \sigma_a(n,2^{z-1}) = 0 \qquad n - 2^{z-2} \text{ different words}$$
$$\sigma_b(n,2^z) = 1 \qquad \sigma_b(n,2^{z-1}) = 1 \qquad n \text{ different w.}$$
$$\sigma_c(n,2^z) = 3 \qquad \sigma_c(n,2^{z-1}) = 1 \qquad 2^{z-1} - n \text{ different w.} \qquad (18)$$

Group Y: $2^{z-2} < n \leq 2^{z-2}+2^{z-3}$ initial conditions:

$$\sigma_a(n,2^z) = 2 \qquad \sigma_a(n,2^{z-1}) = 0 \qquad n - 2^{z-2} \text{ different words}$$
$$\sigma_b(n,2^z) = 1 \qquad \sigma_b(n,2^{z-1}) = 1 \qquad 3n - 2^{z-1} - 2^{z-2} \text{ different w.}$$
$$\sigma_c(n,2^z) = 3 \qquad \sigma_c(n,2^{z-1}) = 1 \qquad 2^{z-1} - n \text{ different w.}$$
$$\sigma_d(n,2^z) = 2 \qquad \sigma_d(n,2^{z-1}) = 2 \qquad 2^{z-2} + 2^{z-3} - n \text{ different w.} \qquad (19)$$

Inserting these initial conditions into formula (17) yields the frequencies of n–words in an arbitrarily long sequence and one can calculate the probability of the words. We get:

Group X: $2^{z-2} + 2^{z-3} \leq n \leq 2^{z-1}$

$$p_a = \frac{4}{3}\frac{1}{2^z} \qquad p_b = \frac{4}{3}\frac{1}{2^z} \qquad p_c = \frac{8}{3}\frac{1}{2^z} \tag{20}$$

Group Y: $2^{z-2} < n \leq 2^{z-2} + 2^{z-3}$

$$p_a = \frac{4}{3}\frac{1}{2^z} \qquad p_b = \frac{4}{3}\frac{1}{2^z} \qquad p_c = \frac{8}{3}\frac{1}{2^z} \qquad p_d = \frac{8}{3}\frac{1}{2^z} \tag{21}$$

We insert the quantities above into formula (1) for the block entropies. The result is: Group X: $2^{z-2} + 2^{z-3} \leq n \leq 2^{z-1}$

$$H_n = (2^{z-2} - 2n)\frac{4}{3*2^z}\log\frac{4}{3*2^z} + (n - 2^{z-1})\frac{8}{3*2^z}\log\frac{8}{3*2^z} \tag{22}$$

Group Y: $2^{z-2} < n \leq 2^{z-2} + 2^{z-3}$

$$H_n = (2^z - 4n)\frac{4}{3*2^z}\log\frac{4}{3*2^z} - (7*2^{z-3} - 2n)\frac{8}{3*2^z}\log\frac{8}{3*2^z} \tag{23}$$

Some rearrangements yield the formulas given in [9].

Figure 4 shows the conditional entropies h_n for $n = 0, 1, 2, \ldots, 29$.

It follows from the arguments above: if the word length is $n = 2^z + 2^{z-1}$ and one goes to the next word length $n = 2^z + 2^{z-1} + 1$ then two new words are excluded. Accordingly at these word lengths the conditional entropy h_n decreases because the average uncertainty about the prediction of a forthcoming symbol decreases. Between these wordlengths the entropy is constant because no new restriction rules occur.

4. Arbitrary Partitions

In the previous sections we only considered symbol sequences resulting from the generating bipartition of state space. In this section we firstly investigate symbol sequences resulting from arbitrary binary partitions and in the sequel partitions related to a larger alphabet.

For this purpose we consider our symbol sequence generator (10): On basis of Lemma 1 one finds in the original binary Feigenbaumsequence repetitions of subsequences of length $2^z - 1$, $z > 0$ and there is one changing symbol, denoted S, between two repetitions. As a consequence of selfsimilarity the S–symbols form the same selfsimilar sequence (besides a permutation of the alphabet).

The S–symbols correspond to a microscopic trajectory in the vicinity of 0.5, either left of 0.5 (symbol 0) or right of 0.5 (symbol 1).

First we consider binary partitions constructed by a backward iterate of 0.5 of arbitrary order:

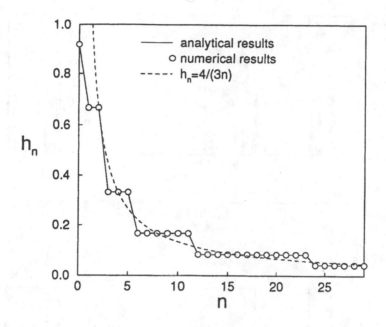

Figure 4. conditional entropy

One can derive from the symbol sequence that only one of the 2^m backward iterates of order m of 0.5 is part of the attractor (for all $m > 0$) and which of the possible backward iterates exists. One has to inquire a long repetitive subsequence and has to read it backwards. The first symbol marks the first backward iterate of 0.5: $1 \rightarrow$ right, the second symbol marks the second backward iterate of 0.5: $0 \rightarrow$ left etc.

It follows that the rules which determine whether a point lies left or right of 0.5 also apply the first, second, etc. backward iterates of 0.5. The sequence generator is the same for these partitions and the resulting sequences are selfsimilar. Only the initial subsequences are changed. In order to exemplify the last statement we compare the binary partition at 0.5 with a binary partition at $f_R^{-1}(0.5)$.

bipartition at 0.5
101110101011101110111010101110101011101010111011101110101011
initial subsequences $a_0 = 1$ and $a_1 = 10$

bipartition at $f_R^{-1}(0.5)$
010001010100010001000101010001010100010101000100010001010100
initial subsequences $a_0 = 0$ and $a_1 = 01$

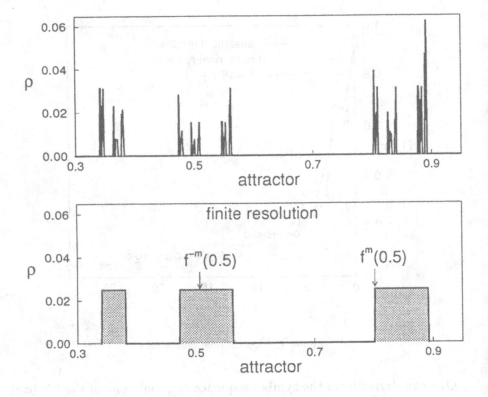

Figure 5.   Feigenbaum accumulation point: histogram

The S–symbols for the partition at $f_R^{-1}(0.5)$ correspond to a microscopic trajectory in the vicinity of $f_R^{-1}(0.5)$.

Due to the Cantor structure of the attractor the support of the invariant measure consists of a fractal set of points. Looking at this set with a finite resolution one can detect a set of intervals which hint at the hidden selfsimilar structure. The backward iterates of 0.5 never can match the edges of the tiny intervals since considering a small section around a backward iterate one will always see the microscopic trajectory on both sides of the backward iterate. Consequently the forward iterates of 0.5 match the edges of the tiny intervals (figure 5). All backward iterates of 0.5 are placed between two forward iterates.

The partition at a forward iterate or at a point which is not part of the attractor leads to a periodic sequence. Considering a small section around the splitting point one will see the microscopic trajectory only on one (forward iterate) or on none (point is not part of the attractor) side of the splitting point. The S–symbols are identical. The symbol sequence is periodic. One finds selfsimilarity on a finite number of scales only.

Next we consider partitions resulting in three or more symbols. We refine the bipartition at 0.5 with splitting points at forward iterates of 0.5. As a first example this was done for a special tripartition choosing 0.5 and $f^3(0.5)$ in [9] . As a result the block entropies are identical with those of the bipartition for $n \geq 3$. We can explain and generalise this result. We observe the following: the probability distributions and the block entropies for a partition at 0.5 and at forward iterates of 0.5 become identical to the values of the bipartition at 0.5 from a certain word length n on (depending on the refinement). The sequence generator is the same for these partitions and the initial subsequences are changed only. We compare the bipartition at 0.5 with a tripartition at 0.5 and at $f^3(0.5)$.

bipartition at 0.5:
10111010101110111011101010111010101110101011101110101011
initial subsequences: $a_0 = 1$ and $a_1 = 10$.

refinement with $f^3(0.5)$:
20212020202120212021202020212020202120202021202120212020202021
initial subsequences: $a_0 = 20$ and $a_1 = 2021$.

The S–symbols are the same for both partitions and the new symbol 2 in the tripartition is on fixed positions in the sequence because the partition at a forward iterate yields a periodic sequence.

For the first few word lengths n the probability distributions differ because the new symbols change the statistics.

Choosing general higher iterates of 0.5 or points which are not part of the attractor in addition to the original splitting point at 0.5 the resulting new symbols will be on fixed positions in the selfsimilar sequence. The reason is that the partition at forward iterates or a points which are not part of the attractor yields a periodic sequence.

Now we study the refinement of the partition at 0.5 with splitting points at backward iterates of arbitrary order of 0.5. As in the case above the symbol sequences can be described by the symbol sequence generator for the bipartition at 0.5 but with different initial subsequences. We compare the bipartition at 0.5 with a tripartition at 0.5 and at $f^{-1}(0.5)$.

bipartition at 0.5:
10111010101110111011101010111010101110101011101110101011
initial subsequences: $a_0 = 1$ and $a_1 = 10$.

refinement with $f_R^{-1}(0.5)$
20112020201120112011202020112020201120202011201120112020201
initial subsequences: $a_0 = 1$ and $a_1 = 20$.

In the tripartition all symbols take part in the game of self similarity. The tripartition is the same as the bipartition at 0.5 for words of length $n = 2$ (in the bipartition at 0.5 the word 00 is forbidden). It occurs a shift of the probability distributions and entropies of the bipartition at 0.5 to lower n.

Generally, choosing l backward iterates of 0.5 – such as $f^{-i_1}(0.5), \ldots,$ $f^{-i_l}(0.5)$ (with $i_j \geq 0$) – all of which are elements of the attractor, the resulting symbol sequence is selfsimilar and the block entropies are shifted to lower n. The reason is that sufficient selfrefinement of this partition corresponds to an appropriate selfrefinement of the binary partition constructed by choosing $f^{-i_*}(0.5)$ where $i_* := \min \{i_1, \ldots i_l\}$.

Collecting all considered cases amounts to controlling an arbitrary partition. Hence, we are able to predict the character of any resulting sequence, i.e. we can classify it as being either periodic or selfsimilar. In the first case the generator is an invariant item of the system; it is only the initial sequences a_0 and a_1 which vary. In fact, even the periodic case can be included in a trivial manner, namely by choosing a_0 to be the basic periodic subsequence and $a_1 = a_0$. This suggests that the idea that the generator reflects features of the microscopic dynamics.

5. Conclusions and Outlook

In this paper we investigated the symbolic dynamics of the logistic map at the Feigenbaum accumulation point. We employed a symbol sequence generator involving deterministic grammatical rules producing the binary selfsimilar Feigenbaumsequence for the partition at $x_{max} = 0.5$. We analytically derived the block entropies for all word lengths n on basis of these grammatical rules. The result is a power law decay of the conditional entropies $h_n \sim \frac{1}{n}$ corresponding to an infinite memory. The grammatical rules allowed an extension of results to arbitrary binary partitions and in the sequel to arbitrary partitions. We have shown, that the resulting symbol sequences are either periodic or selfsimilar.

Referring to the introduction we may consider a possible connection between the Feigenbaumsequence and natural sequences like text and music. In [21, 27] we introduced a nonlinear dynamical model for texts. The model contains deterministic and stochastic rules. The conditional entropies show a power law decay $h_n \sim 1/n^\alpha$ with $\alpha > 0$ and where $\frac{1}{1-\alpha}$ is an integer.

There exist similarities between the symbol sequence generator producing the Feigenbaumsequence and the nonlinear dynamical model for texts. Both give rise to resp. work with repetitions of subsequences. Between two such repetitions the Feigenbaumsequence assigns a fully determined symbol whereas the nonlinear dynamical model for texts allows for a stochastic selection. Obviously, natural sequences are not perfectly selfsimilar. But a simple perturbation of a selfsimilar sequence by flipping each symbol with the probability ϵ does not change the scaling law of conditional entropies $h_n \sim 1/n$ [17, 26]. Perhaps a more sophisticated mechanism of perturbing selfsimilar sequences might lead to a power law with $\alpha < 1$ and moreover, might establish contact to the nonlinear model for texts.

References

1. Broer, H.W. (1991) in E. van Groesen and E.M. de Jager (eds.) *Structures in Dynamics–Finite Dimensional Deterministic Studies*, North–Holland/Elsevier Science, Amsterdam, pp. 1–23.
2. Bai–lin, H. (1984) *Chaos*, World Scientific, Singapore.
3. Collet, P. and Eckmann, J.P. (1980) *Iterated Maps of the Interval as Dynamical Systems*, Birkhäuser, Basel.
4. Alekseev, V.M. and Yakobson, M.V. (1981) *Phys. Rep.* **75**, 290–325.
5. Arnold,V.I. and Avez, A. (1968) *Ergodic Problems of Classical Mechanics*, Benjamin, New York.
6. Eckmann, J.-P. and Ruelle, D. (1985) *Rev. Mod. Phys.* **57**, 617–656.
7. Ebeling, W., Engel, H. and Herzel, H. (1990) *Selbstorganisation in der Zeit*, Akademie–Verlag, Berlin.
8. Ebeling, W. and Nicolis, G. (1991) *Europhys. Lett.* **14**, 191– 196.
9. Ebeling, W. and Nicolis, G. (1992) *Chaos, Solitons & Fractals* **2**, 635–650.
10. Szépfalusy, P. (1989) *Physica Scripta* T **25**, 226–229.
11. Khinchin, A.I. (1957) *Mathematical Foundations of Information Theory*, Dover Publisher, New York.
12. McMillan, B. (1953) *Ann. Math. Statist.* **24**, 196–216.
13. Schuster, H.G. (1989) *Deterministic Chaos: An Introduction*, VCH Verlag, Weinheim.
14. Szépfalusy, P. and Györgyi, G. (1986) *Phys. Rev.* **A33**, 2852–2855.
15. Grassberger, P. (1986) *Intern. Journ. Theo. Phys.* **25**, 907– 938.
16. Gramss, T. (1995) *Phys. Rev.* E **50**, 2616–2620.
17. Freund, J., Ebeling, W. and Rateitschak, K. (1996) submitted to *Phys. Rev.* **E**.
18. Ebeling, W. and Pöschel, T. (1994) *Europhys. Lett.* **26**, 241–246.
19. Ebeling, W., Pöschel, T. and Albrecht, K.F. (1995) *Intern. Journ. for Bifurc. & Chaos* **5**, 51–61.
20. Ebeling, W., Neiman, A. and Pöschel, T. (1995) in: M. Suzuki (ed.) *Coherent Approach to Fluctuations (Proc. Hayashibara Forum 95)*. World Scientific, Singapore.
21. Rateitschak, K., Ebeling, W. and Freund, J. (1996) submitted to *Phys. Rev.* **E**.
22. Herzel, H. and Große, I. (1995) *Physica* A **216**, 518–542.
23. Herzel, H., Ebeling, W. and Schmitt, A.O. (1995) *Phys. Rev.* E **50**, 5061–5071.
24. Hopcroft, J.E. and Ullman, J.D. (1979) *Introduction to Automata Theory, Languages and Computation*, Addison–Wesley, Reading.
25. Feigenbaum, M.J. (1978) *J. Stat. Phys.* **19**, 25–52.
26. Freund, J. (1995) *Dissertation*, Humboldt University Berlin.
27. Rateitschak, K. (1995) *diploma thesis*, Humboldt University Berlin.

28. Schroeder, M.R. (1991) *Fractals, Chaos, Power Laws*, Freeman, New York.
29. Li, W. (1991) *Phys. Rev.* **A 43**, 5240–5260.
30. Li, W. (1992) *Int. Journ. for Bifurc. & Chaos* **2**, 137–154.
31. Jaglom, A.M. and Jaglom, I.M. (1984) *Wahrscheinlichkeit und Information*, VEB Deutscher Verlag der Wissenschaften, Berlin.
32. Kolmogorov, A.N. (1965) *Probl. of Inform. Theory* **1**, 3–13.
33. Chaitin, G.J. (1975) *Journal of the A. C. M.* **22**, 329–345.
34. Grassberger, P. (1989) *Helv. Phys. Acta* **62**, 489–508.
35. Atmanspacher, H., Kurths, J., Scheingraber, H., Wackerbauer, R. and Witt, A. (1992) *Open Systems & Information Dynamics* **1**, 269–276.
36. Wackerbauer, R., Witt, A., Atmanspacher, H., Feudel, F., Kurths, J. and Scheingraber, H. (1994) *Chaos, Solitons & Fractals* **4**, 133–173.

DEVELOPMENTS IN NORMAL AND GRAVITATIONAL THERMODYNAMICS

GEOFFREY L. SEWELL
Department of Physics, Queen Mary and Westfield College
Mile End Road, London E1 4NS, U.K.

Abstract.

We review developments in three areas of thermodynamics, resulting from advances in statistical mechanics and relativity theory. These concern (a) the resolution of some basic questions, concerning the thermodynamic variables and the phase structure of normal matter, (b) thermodynamical instabilities in non-relativistic gravitational systems, and (c) Black Hole thermodynamics, as formulated in terms of strictly observable quantities, and thus not involving any BH entropy concept.

1. Introduction

The object of this article is to discuss advances in three areas of thermodynamics, that have stemmed from progress over the last three decades in quantum statistical mechanics and general relativity. I shall keep the mathematics here very simple, even though the advances that I am going to discuss depend largely on results obtained by rather abstract arguments.

The three areas of thermodynamics I shall discuss are those of normal systems, i.e. ones whose energies are extensive variables; non-relativistic gravitational systems, whose energies are not extensive, because of the long range and attractivity of the Newtonian interactions; and Black Holes, which arise as a consequence of relativistic gravitational collapse of stars.

I shall start, in §2, with an expose' of developments in the statistical thermodynamics of normal systems, that have been achieved within the framework in which a macroscopic system is represented as an infinitely extended one of finite density [1-3]. This amounts to an idealisation, which serves to reveal intrinsic bulk properties of macro-systems, that are oth-

27

J. S. Shiner (ed.), Entropy and Entropy Generation, 27–36.
© *1996 Kluwer Academic Publishers.*

erwise masked by boundary and other finite-size effects. In particular, by contrast with the traditional statistical mechanics of finite systems, it accommodates the *phase structure* of matter, as manifested not only by the singularities in thermodynamic potentials, but also by the coexistence of equilibrium states with different microstructures. Here, I shall sketch two basic and relatively new results, obtained within this framework [3, Ch.4], that go beyond those of traditional statistical, as well as classical, thermodynamics. The first provides an answer to the fundamental question of what comprises a complete set of macroscopic observables, corresponding to the thermodynamic variables of a system. The second establishes a statistical mechanical basis for the empirically known connection between two *a priori* distinct characteristics of phase transitions, namely thermodynamic singularities and phase coexistence.

I shall then provide a brief note, in §3, on the thermodynamics of non-relativistic gravitational systems, consisting of fermions of a single species that interact via Newtonian forces. Here, the main result [4] is that these systems have negative *microcanonical* specific heat in a certain regime, and consequently are unstable there with respect to energy exchanges with thermal reservoirs.

In §4, I shall pass to the thermodynamics of Black Holes. This poses rather special conceptual problems because, by its very nature, a Black Hole has no observable microstructure. Hence, the standard concept of entropy as a structural property of an operationally determinable microstate [5, Ch.5] is inapplicable to it; and therefore the well-known Bekenstein-Hawking proposal [6,7], which is designed to extend this concept to Black Holes, is defective because of its subjective component. In order to restore objectivity to the theory, I have made a different approach to the thermodynamics of processes involving the exchange of matter with Black Holes [8]. This is based exclusively on observable quantities and thus involves no BH entropy concept. Here, I shall provide a simple treatment of the argument leading to Bekenstein's formula [6] for a generalised second law of thermodynamics, but now interpreted within a framework built on the observables of the exterior region of the Black Hole.

2. Normal Systems

A normal system is one whose energy is an extensive variable. This means that the energy of interaction between the particles in two disjoint regions is a 'surface effect'. At the level of atomic physics, this extensivity condition is fulfilled by systems with interactions possessing realistic properties of short range and stability [1]. At a more microscopic level, it is also satisfied by electrically neutral systems of electrons and nuclei, that interact via

Coulomb forces [9]. Here the roles of both the Fermi statistics and the Debye screening are crucial.

As is well known, the extensivity property permits one to represent normal macroscopic systems in the so-called thermodynamical limit, where they are idealised as infinitely extended assemblies of particles [1-3]. This is the model we shall employ, since it is needed for the mathematically sharp characterisation of macroscopic phenomena such as thermodynamic singularities, phase coexistence and irreversibilty.

2.1. THERMODYNAMIC VARIABLES

In Classical Thermodynamics (CT), the macrostate of a system is represented by a set $q = (q_1, .. , q_n)$ of intensive variables, that are global densites of extensive conserved quantities, such as energy, magnetic moment, etc. The entropy density is then a function, s, of q, whose form governs the thermodynamics of the system. We note here that, for any given system, *CT provides no specification of the variables, q, or even of their number, n.* For that, as well as for the form of s, we need to pass to the statistical mechanical model, Σ, of the system.

We shall now sketch a statistical thermodynamical scheme we have devised for the purpose of characterising the variables q and formulating the function s in terms of the microscopic structure of the system [3, Ch.4]. For this, we introduce the key concept of *thermodynamic completeness*, as applied to a set $\hat{Q} = (\hat{Q}_1, .. , \hat{Q}_n)$, of extensive conserved observables of Σ, in the following way. Denoting by $\hat{q} = (\hat{q}_1, .. , \hat{q}_n)$ the intensive observables, given by the global densities of $\hat{Q}$, we term $\hat{Q}$ thermodynamically complete if

(C.1) $\hat{Q}_1$ *is the energy;*
(C.2) $(\hat{Q}_1, .. , \hat{Q}_n)$ *are linearly independent; and*
(C.3) *if $\hat{s}(\rho)$ is the entropy density[1] of a translationally invariant mixed state ρ of Σ, then*
(a) *for any given expectation value q of $\hat{q}$, there is precisely one state, ρ_q, that maximises $\hat{s}$, and*
(b) *the same is not true for any proper subset of $(\hat{q}_1, .. , \hat{q}_n)$.*

In other words, $\hat{Q}$ is thermodynamically complete if the value of its density, $\hat{q}$, determines the equilibrium microstate, without redundancy. For example, in the case of a ferromagnetic system, such as the Ising model, $\hat{q}$ comprises the energy density and the polarisation.

---

[1]This is defined [1, Ch.7] as a natural generalisation, to infinite systems, of the density of the Von Neumann entropy, $-kTr(\rho\log\rho)$, of a microstate whose density matrix is ρ.

We assume, as the condition for Σ to support a thermodynamics, that it possesses a complete set of extensive conserved observables, $\hat{Q}$, that is unique, up to linear combinations. The intensive thermodynamical variables of the system are then the expectation values, q, of the global densities, $\hat{q}$, of $\hat{Q}$.

2.2. THERMODYNAMIC POTENTIALS

The equilibrium entropy density of Σ, as a function of the thermodynamical variables q may be seen from the condition (C.3a) to be

$$s(q) \equiv \hat{s}(\rho_q) \tag{2.1}$$

In fact, this is a generalisation of the standard definition of microcanonical entropy [10], and reduces to that in the case where $\hat{Q}$ consists of the energy only. Further, the potential s has the following key properties.

(S.1) s is concave [3, Ch.4], i.e.

$$s(\lambda q + (1 - \lambda)q') \geq \lambda s(q) + (1 - \lambda)s(q') \tag{2.2}$$

which represents the thermodynamical stability of the system.

(S.2) s is differentiable [11, Appendix A]. This implies that that the inverse temperature, $\partial s/\partial q_j$, is always well-defined.

We can also describe the thermodynamics of Σ in terms of the variables $\theta = (\theta_1, \ldots, \theta_n)$, conjugate to q and defined as the partial derivatives of s w.r.t. $(q_1, \ldots, q_n)$, respectively. The thermodynamic potential appropriate to this description is

$$p(\theta) = s(q) - q.\theta, \text{ with } \theta_j = \frac{\partial s}{\partial q_j} \tag{2.3}$$

where $q.\theta \equiv \sum_{j=1}^{n} q_j \theta_j$. We note that $p(\theta)$ is just the product of the pressure and the inverse temperature, θ_1 [1, P.57]; and further, in view of the concavity of s, our above definition of it is equivalent to the formula

$$p(\theta) = \max_q (s(q) - \theta.q) \tag{2.3}'$$

2.3. COMMENTS

(1) It follows from this last equation that p is convex [3, Ch.4], i.e.

$$p(\lambda\theta + (1 - \lambda)\theta') \leq \lambda p(\theta) + (1 - \lambda)p(\theta') \tag{2.4}$$

This formula, like (2.2), represents the thermodynamical stability of the system. At the mathematical level, it implies [12] that the function p is continuous, except possibly at the boundaries of its domain of definition. On the other hand, it does not guarantee the differentiability of p. In fact, this function can support singularities, as given by points, θ, where it is not differentiable [1,3]. Such points correspond, of course, to phase transitions. (2) The maximisation of $s - \theta.q$ corresponds to the minimisation of the Gibbs free energy and thus occurs precisely when q takes an equilibrium value, $\bar{q}$. By the completeness condition (C3), this determines the corresponding equilibrium microstate $\rho_{\bar{q}}$. Thus, the condition for *phase coexistence*, for given θ, is that $s - \theta.q$ attains its maximum at more than one value of q.

2.4. EQUILIBRIUM STATES AND PHASE STRUCTURE

This last observation brings us to the point that, in Classical Thermodynamics, there are two characterisations of phase transitions, namely,

(a) singularities in thermodynamic potentials; and
(b) coexistence of different equilibrium states.

In fact, although (a) and (b) are *a priori* distinct from one another, it is generally assumed, on empirical grounds, that they always arise together. It is therefore natural to ask whether there is any *neccessary* connection, imposed by the underlying quantum statistical structure, between these characterisations. The answer to this question is provided by the following Proposition, which was proved [3, Ch.4] on the basis of the concavity/convexity properties (2.2) and (2.4).

2.5. PROPOSITION

Under the above definitions and assumptions, phase coexistence occurs at precisely those values of θ where the potential p is not differentiable.

In other words, the structures imposed by the quantum statistical model render the characterisations (a) and (b) equivalent.

3. Non-relativistic Gravitational Systems

We consider now a system, Σ_N, of N electrically neutral, massive fermions of one species, interacting via Newtonian forces. This is a model of a neutron star, and its Hamiltonian takes the form

$$H_N = -\frac{\hbar^2}{2m}\sum_{j=1}^{N}\Delta_j - \kappa m^2 \sum_{j,k(\neq j)=1}^{N} r_{jk}^{-1} \qquad (3.1)$$

where κ is the gravitational constant, Δ_j is the Laplacian for the j'th particle, and r_{jk} is the distance between the j'th and k'th particles.

Because of the long range and attractivity of the interactions, the energy of the system is *not* an extensive variable, and so the theory of §2 is inapplicable to this model. In fact [4], for large N, the volume occupied by Σ_N in equilibrium is proportional to N^{-1}, rather than N. This result represents the outcome of the competition between the Newtonian attraction, which favours implosion, and the Fermi pressure. Furthermore, the energy, E_N, and the microcanonical entropy, S_N, are proportional to $N^{7/3}$ and N, respectively.

Thus, the statistical thermodynamics of the model may be formulated according to a scheme [4] in which, for each value of N, the system Σ_N is confined to a spatial region Ω_N, such that

(a) for different values of N, the regions Ω_N are geometrically similar, i.e. they all have the same shape; and

(b) the volume, V_N, of Ω_N is proportional to N^{-1}.

The thermodynamic properties of the model that emerge from this scheme are the following [4].

(G1) *If $N^{-7/3}E_N$, NV_N converge to e, v, respectively, as $N\to\infty$, then the microcanonical specific entropy $N^{-1}S_N$ converges to a function, s, of (e, v).*

(G2) *There is a regime in which s is convex w.r.t. e, i.e. where the system is unstable against energy exchanges with a thermal reservoir.*

(G3) *The system undergoes a phase transition, of the Van der Waals type, when s changes from concave to convex w.r.t. e.*

For further properties of the model, see Refs. [2, Section 4.2] and [13-15]. In particular, when N becomes sufficiently large, $\approx 10^{60}$, the non-relativistic model becomes unphysical, since it implies that the mean particle velocities are comparable with the speed of light.

4. Black Hole Thermodynamics

According to Classical General Relativity, the collapse of a star, due to gravitational implosion, results in the formation of a Black Hole [16, Pt.7], i.e. a bounded spatial region, from which no light can escape. Thus, the only observables of the BH that can be perceived from the outside are its mass, electric charge and angular momentum, these being the ones that are registered by the external gravitational and electromagnetic fields [16, P.876]. This is Wheeler's famous principle that "Black Holes have no Hair", the word 'hair' here meaning 'microstructure that can be observed from outside'. It has dramatic consequences for thermodynamics, since it implies that a BH has no operationally determinable microstate. The concepts of statistical mechanics, including that of entropy, are therefore inapplicable to it, and, consequently, the standard form of the Second law of Thermodynamics has no predictive value for processes in which observable systems discharge matter into the Hole.[2]

On the other hand, the relativistic mechanics of Black Holes [17] exhibits remarkable analogies with classical thermodynamics, with the surface area of a Hole playing the role of entropy. Bekenstein [6] argued that these analogies carried physical content by showing that, in certain Gedankenexperiments, in which matter was dropped into a Black Hole, the sum of the entropy, S, of the exterior region and the area, A, of the BH, multiplied by a certain constant, λ, never decreased, i.e.,

$$\Delta S + \lambda \Delta A \geq 0, \tag{4.1}$$

On this basis, he proposed that the BH had entropy λA, and that processes involving BH's conformed to a Generalised Second Law (GSL), represented by equn. (4.1). This proposal was supported [7,18] by Hawking's subsequent argument that Black Holes act as sources of thermal radiation, of quantum mechanical origin, with temperature corresponding to the assumed form, λA, of the BH entropy. At the quantum statistical level, Bekenstein introduced a *subjective* element into the theory, suggesting that this entropy was an information-theoretic quantity, that represented the external observer's ignorance of the state of the BH.

In order to recast the theory in purely objective terms, that retain the standard connections between thermodynamic variables and underlying microstructures, we have provided a different treatment of processes involving Black Holes [8]. This is based on the statistical thermodynamics of the exterior, observable region, and does not involve any concept of a BH entropy. As we shall see, it leads to a version of the GSL (4.1), as adapted to open systems, but with the last term there representing mechanical work on the BH, not entropy. We emphasise here that, as in the previous works, our treatment is based on the model in which the gravitational field and the geometry are classical, and therefore where the "No Hair" Principle is applicable.

We formulate the theory on the basis of a thermodynamics of a testbody, Σ, which is placed in the Hawking radiation of a Black Hole, B, and exchanges mass, electric charge and angular momentum with both B and the radiation. Σ is thus an *open system*. Accordingly, we base our treatment on the following general principles.

(I) *The Second Law of Thermodynamics, which tells us that the Gibbs potential, Φ, of an open system cannot spontaneously increase. Equivalently*

---

[2]This remark was credited to J. A. Wheeler by Bekenstein [6].

[19, Ch.2], if mechanical work, W, is done on the system, then

$$W \geq \Delta\Phi \qquad (4.2)$$

(II) *The laws of Black Hole Mechanics [16], which ensue from General Relativity. These are that*
(BH1) the total energy, electric charge and angular momentum of the BH and the exterior system are conserved in any process;
(BH2) the surface gravity, σ, i.e. the (proper) acceleration of a freely infalling body at the surface of B, is uniform over that surface;
(BH3) the surface area, A, of B is a function of its energy, E_B, charge, Q_B, and angular momentum, J_B, and satisfies the differential relation

$$\sigma c^2 dA/8\pi\kappa = dE_B - \phi dQ_B - \omega.dJ_B \qquad (4.3)$$

where ϕ is the electric potential and ω the angular velocity of B; and
(BH4) A increases in irreversible processes, and remains constant in reversible ones. Here, it is the the origination of the BH from the collapse of a star that is the source of the irreversibility.
(III) *The Hawking thermal radiation phenomenon [18], which we have shown [20] to be a general, model-independent consequence of the basic principles of quantum theory, relativity and statistical thermodynamics. The temperature of this radiation is*

$$T = \hbar\sigma/2\pi kc \qquad (4.4)$$

and its electric potential and angular velocity are those of B, namely ϕ and ω, respectively.

4.1. NOTE

The laws of Black Hole Mechanics, (BH1-4), have an obvious analogy with classical thermodynamics, with σ and A playing the roles of temperature and entropy, respectively. However, we do *not* interpret A as an entropy, for the following reasons.

(1) In the case of a normal physical system, a microstate corresponds to a density matrix, ρ, whose explicit form may be operationally determined, and the entropy is a function, $-kTr(\rho\log\rho)$, of this state, which provides a measure of its disorder [2, P.57]. Thus, its essential significance is not merely that it has an information-theoretic form, but that it represents a *structural property* of the microstate. In the case of a Black Hole, however, no similar quantity could exist, since the 'No Hair' Principle rules out the possibility of its having an observable microstructure.
(2) One sees immediately from (4.3) that the last term in (4.1) l.h.s. corresponds to the mechanical work required to effect changes ΔE_B, ΔQ_B, ΔJ_B

in E_B, Q_B, J_B, respectively. Hence, apart from the considerations of (1), there is no call for it to be regarded as an entropy.

(3) Whereas thermodynamic irreversibility arises as a result of 'phase mixing', there is no such mechanism operating in (BH4).

4.2. DERIVATION OF THE GSL

We consider the thermodynamics of a test-body, Σ, placed in the Hawking radiation in the asymptotically flat region outside a Black Hole, B. Thus, by (III), Σ is immersed in a heat bath of temperature T, electric potential ϕ and angular velocity ω. Its Gibbs potential is therefore

$$\Phi = E - TS - \phi Q - \omega.J \tag{4.5}$$

where E, S, Q and J are its energy, entropy, charge and angular momentum, respectively. Thus, in any process in which work ΔW is done on Σ by some external device, the resultant changes in Φ, E, S, Q and J must satisfy the following demand of the classical Second Law of Thermodynamics.

$$\Delta W \geq \Delta\Phi = \Delta E - T\Delta S - \phi\Delta Q - \omega.\Delta J \tag{4.6}$$

Suppose now that the process results in an exchange of matter between Σ and B. Then it follows from the conservation laws ($BH1$) that

$$\Delta W = \Delta E + \Delta E_B; \quad \Delta Q + \Delta Q_B = 0; \ and \ \Delta J + \Delta J_B = 0 \tag{4.7}$$

Hence, by equns. (4.6) and (4.7),

$$T\Delta S + \Delta E_B - \phi\Delta Q_B - \omega.\Delta J_B \geq 0 \tag{4.8}$$

i.e., by equns. (4.3) and (4.4),

$$\Delta S + \lambda\Delta A \geq 0; \ with \ \lambda = kc^3/4\kappa\hbar \tag{4.9}$$

This is precisely the GSL proposed by Bekenstein, though now we interpret the area term as mechanical work, stemming from a surface energy, rather than entropy (cf. Note (2) following equn. (4.4)).

Finally, we note that this derivation of the GSL may easily be extended to situations where some of the energy, etc, of Σ is transferred to the Hawking radiation field. For, in that case, the equns. (4.7)-(4.9) remain valid if ΔS, ΔE, ΔQ and ΔJ are interpreted as the total changes of entropy, energy, charge and angular momentum of the matter and radiation in the exterior region.

Acknowledgement

This work was partially supported by European Capital and Mobility Contract No. CHRX- Ct.92-0007.

References

1. Ruelle. D. (1969) *Statistical Mechanics*, W. A. Benjamin, Inc., New York.
2. Thirring, W. (1980) *Quantum Mechanics of Large Systems*, Springer, New York.
3. Sewell, G.L. (1989) *Quantum Theory of Collective Phenomena*, Oxford University Press, Oxford.
4. Hertel, P. and Thirring, W. (1971) in H.P. Durr (ed.) *Quanten und Felder*, Vieweg, Braunschweig, pp. 310-323.
5. Von Neumann, J. (1955) *Mathematical Foundations of Quantum Mechanics*, Princeton University Press, Princeton.
6. Bekenstein, J. (1973) *Phys. Rev. D* **7**, 2333-2346; and (1981) in Y. Neeman (ed.) *Jerusalem Einstein Centenary Conference*, Addison-Wesley, Reading, pp. 42-62.
7. Hawking, S.W. (1976) *Phys. Rev. D* **13**, 191-197.
8. Sewell, G.L. (1987) *Phys. Lett. A* **122**, 309-311; (1987) *Phys. Lett. A* **123**, 499 (Erratum).
9. Lieb, E.H. and Lebowitz, J.L. (1972) *Adv. Math.* **9**, 316-398.
10. Griffiths, R.B. (1965) *J. Math. Phys.* **1447**, 1447-1461.
11. Sewell, G.L. (1991) in W. Gans, A. Blumen and A. Amann (eds.) *Large-Scale Molecular Systems: Quantum and Stochastic Aspects* (*Nato ASI Series B*), Plenum, New York, pp. 77-122.
12. Rockafellar, R.T. (1970) *Convex Analysis*, Princeton University Press, Princeton.
13. Hertel, P., Narnhofer, H. and Thirring, W. (1972) *Commun. Math. Phys.* **28**, 159-176.
14. Narnhofer, H. and Sewell, G.L. (1980) *Commun. Math. Phys.* **71**, 1-28.
15. Messer, J. (1981) *J. Math. Phys.* **22**, 2910-2917.
16. Misner, C.W., Thorne, K. and Wheeler, J.A. (1973) *Gravitation*, W. H. Freeman, San Francisco.
17. Bardeen, J.M., Carter, B. and Hawking, S.W. (1973) *Commun. Math. Phys.* **31**, 161-170.
18. Hawking, S.W. (1975) *Commun. Math. Phys.* **43**, 199-220.
19. Landau, L.D. and Lifschitz, E.M. (1959) *Statistical Physics*, Pergamon Press, Oxford.
20. Sewell, G.L. (1982) *Ann. Phys.* **41**, 201-224.

EXTENDED IRREVERSIBLE THERMODYNAMICS: STATEMENTS AND PROSPECTS

G. LEBON
Université de Liège, Institut de Physique B5, Sart Tilman
B-4000 Liège 1, Belgium

AND

D. JOU AND J. CASAS-VÁZQUEZ
Universitat Autònoma de Bareclona, Departament de Física
E-08193 Bellaterra, Barcelona, Spain

Abstract. Extended irreversible thermodynamics has known a wide development during the two last decades. In this work, the main statements of the theory are reviewed and a panoramic view of some of the open problems to be studied in the future years is presented.

1. Introduction

Nonequilibrium thermodynamics has many faces: the most popular theory, referred to as Classical Irreversible Thermodynamics (CIT), was developed among others by Meixner, Onsager, and Prigogine. A more general but much more formal approach –the so-called rational thermodynamics– was proposed by Coleman, Noll, and Truesdell. Extended Irreversible Thermodynamics (EIT) is known as a new formalism which has fuelled much interest during the last decade. In the present work, basic statements underlying this formalism are reviewed and a list of open questions, problems and prospects are examined.

EIT has been a very active branch of phenomenological nonequilibrium thermodynamics during the last fifteen years. Several proceedings and monographs [1-7], some reviews [8-12] and more than four hundred papers have been published on this topic.

EIT provides a mesoscopic and causal description of nonequilibrium processes: it was originally born out of the double necessity to go beyond the hypothesis of local equilibrium and to avoid the paradox of propagation of disturbances with an infinite speed. This physically unpleasant property arises as a consequence of the substitution of the Fourier, Newton and Fick laws in the balance equations of energy, momentum and mass respectively. Indeed, the resulting partial differential equations are parabolic from which follows that disturbances will be felt instantaneously in the whole space. In most practical situations, the problem of infinite speed of propagation is not relevant as those parts of the signal having infinite velocity are strongly damped at room temperature.

37

J. S. Shiner (ed.), Entropy and Entropy Generation, 37–54.

This is no longer true at low temperature where damping may become unimportant.

The treatment of a thermodynamic system, say a gas of N particles ($N \approx 10^{23}$), can be performed at three different levels of description. At the microscopic level, the description of each particle demands six variables (three for position, three for velocity) for each particle so that in total one needs $6N$ variables. At the macroscopic level, in hydrodynamics, only five variables (e.g. the mass density, the three components of the barycentric velocity, and the temperature) are necessary, the gap between both formalisms being bridged by statistical mechanics. But some systems or some physical situations may require more variables than five and less than $6N$: EIT falls precisely in such an intermediate mesoscopic description.

Though modifications of the classical transport equations leading to finite speed of propagation were known since Maxwell (1865) and especially worked out by Cattaneo (1947), Vernotte (1958), and Grad (1949) among others, the thermodynamic implications did not receive almost any attention until the early works of Nettleton (1959) and Müller (1967) although they were forgotten during more than a decade. It was at the end of seventies when a true explosion of interest on these topics inside several independent groups happened, and thence it has pursued growing up to lead to a systematic, but plural, approach to nonequilibrium thermodynamics [2,3,7]. Like in any macroscopic formalism, the fundamental problem is from one side, the identification of the set of relevant variables and, from the other side, the formulation of evolution equations for these variables. Moreover, one has to take into account restrictions placed by the general laws of physics, like the second law of thermodynamics.

The aim of this paper is mainly to provide a short account of the state of the art and to show a prospective view underlying some of the topics which may focus an active attention in the near future and foster a spread of the theory amongst practitioners of nonequilibrium thermodynamics.

2. The Basic Statements of EIT

EIT is not an unique theory but rather a general way to approach a wide set of situations which are beyond the local-equilibrium hypothesis: there are several versions based on different points of view and perspectives. The present section will concern the main features underlying EIT, which may be summarized as follows.

2.1. ENLARGEMENT OF THE SPACE OF INDEPENDENT VARIABLES

In the classical description, one uses generally five variables to describe a fluid, i.e. the mass density ρ, the three components of the velocity vector v and the internal energy per unit mass u (or temperature T). These variables will be now referred to as the subset C of classical variables. In EIT one includes in addition a new subset of variables, denoted by F, which are formed by physical fluxes (of mass, momentum, heat) and higher-order fluxes associated to the variables in C, as well as others deemed necessary for a more complete description of the system under study. The space of state variables, to be denoted by V, consists of the union of the two subsets C and F

$$V = C \cup F$$

The physical nature of the F variables is completely different from that of the C variables. The latter are conserved and slow, as their behaviour is governed by conservation laws, and their slow decay makes them easily accessible to experiments. In contrast, the F variables are non-conserved and fast: they do generally not satisfy conservation laws and their rate of decay may be very fast. In dilute gases, it is of the order of the collision time τ_c between the molecules, i.e. 10^{-10} s. This means that for time intervals much longer than τ_c, the rate of variation of the fast variables can be ignored. This is of course no longer true at very high frequencies ω comparable to $1/\tau_c$ and in systems characterized by a time of decay of the fluxes comparable to the time-scale of the experiments.

Examples of fast phenomena are: ultrasound propagation and light scattering in gases, neutron scattering in liquids, fast explosions or implosions, as those found in supernovae or in laser-induced fusion, decoupling periods in the expanding universe, shock waves, nuclear collisions, transport processes in microelectronic devices of submicronic size. Examples of systems with long relaxation times are: heat conducting solids at low temperature ($\tau_c \approx 10^{-5}$ s), superconductors and superfluids ($\tau_c \approx 10^6$ s), suspensions and polymeric solutions ($\tau_c \approx 10^2$-10^3 s), rarefied gases ($\tau_c \approx 10^2$ s).

2.2. EVOLUTION EQUATIONS FOR THE C AND F VARIABLES

The classical variables C satisfy the usual conservation laws of mass, momentum and energy: they are well known and given by

$$\dot{\rho} = -\rho \nabla.v, \tag{1}$$

$$\rho \dot{v} = -\nabla.P + \rho f, \tag{2}$$

$$\rho \dot{u} = -\nabla.q - P : \nabla v. \tag{3}$$

The notation is classical: an upper dot stands for the material time derivative, v is the barycentric velocity, f the body force per unit mass, q the heat flux, and P the pressure tensor supposed symmetric

$$P = pI + P^v,$$

with p the hydrostatic pressure, I the unit tensor and P^v the viscous pressure tensor. Now, concerning the flux variables F, we have a priori no information about their behaviour in the course of time and space. By analogy with the classical variables, it is assumed that these variables obey first-order time evolution equations of the following general form which is reminiscent of the classical balance equations (1-3):

$$\rho \dot{J} = -\nabla.J^F + \sigma^F. \tag{4}$$

J designates any flux (for instance the flux of matter, momentum, heat, ...), J^F and σ^F denote the corresponding flux and source term, respectively. To require that all the variables of the system obey first-order differential equations means that the behaviour of the system in the space V is local in time, i.e. Markovian; thus, if the behaviour is non-Markovian in the space C, fast variables should be added to V until the behaviour becomes Markovian.

At this stage of the analysis, J^F and σ^F are unknown quantities: they must be expressed in terms of the whole set V of variables by means of constitutive relations of the form $J^F = J^F(V)$, $\sigma^F = \sigma^F(V)$. The explicit form of these constitutive equations will be derived from the representation theorems of tensors and can be formulated at any order of approximation. Typical examples will be analysed in the next section.

The important difference with the classical theory of irreversible processes is that now the heat flux and the pressure tensor are not given by means of a constitutive equation such as

$$q = -\lambda\nabla T \quad \text{(Fourier's law)}, \quad \mathbf{P}^v = -2\eta(\nabla v)^{sym} \quad \text{(Newton's law)} \qquad (5)$$

but are viewed as independent quantities at the same footing as ρ, v and u. The simplest examples of the evolution equations for the fluxes are

$$\tau_1\dot{q} + q = -\lambda\nabla T, \qquad (6)$$

$$\tau_2\dot{\mathbf{P}}^v + \mathbf{P}^v = -2\eta(\nabla v)^{sym}. \qquad (7)$$

These equations generalize Fourier and Newton laws of classical thermomechanics by incorporating relaxational terms, characterized by relaxation times τ_1 and τ_2.

2.3. EXISTENCE OF A NONEQUILIBRIUM ENTROPY

It is assumed that there exists a specific entropy s, which is function of the whole set V of variables, and subject to the following properties:

i. s is an extensive quantity;

ii. s is a convex function of the state variables;

iii. its rate of production σ^s is positive definite, σ^s being defined by

$$\sigma^s = \rho\dot{s} + \nabla.\mathbf{J}^s, \qquad (8)$$

with $\mathbf{J}^s$ the entropy flux which, like s, is supposed to be dependent of the whole set of variables.

Evolution and constitutive equations have to comply with the restrictions imposed by the second law and the convexity requirement of s. The entropy plays a decisive role in the kinetic and statistical interpretations of the foundations of EIT, and therefore we shall come back to this point later on.

Let us now discuss in more details the implications of the above three statements. In view of statement (*i*), entropy is a field variable that is defined at each position and at each instant of time. The second statement (*ii*) is introduced to recover the classical property that equilibrium is a stable state of maximum entropy; in mathematical form, it can be expressed as

$$\left(\frac{\partial^2 s}{\partial V_i \partial V_j}\right)_{eq} < 0 , \tag{9}$$

in which subscript *eq* refers to the equilibrium state and V_i stands for any variable of the set V. Inequality (9) is important because it leads to the conclusion that the relaxation times are positive quantities. This ensures that the set of evolution equations is hyperbolic, allowing the disturbances to propagate with a finite velocity, even at infinite frequencies. This result is at variance with the corresponding results of CIT, wherein the momentum equation and the diffusion equations of temperature and mass are of the parabolic type, with as consequence that the disturbances will propagate at infinite velocity. Finally the statement (*iii*) allows us to determine the most general expressions of the evolution equations for the fluxes which are compatible with the second law of thermodynamics.

EIT reduces to CIT when the relaxation times of the fluxes vanish, and clearly goes beyond this theory by including the physical fluxes and other quantities (subset F) among the set of independent variables and by formulating the second law of thermodynamics in terms of a generalized entropy which depends on these F variables, besides the classical variables belonging to subset C. Such an extension may be very useful to better comprehend the limits of validity of the local equilibrium hypothesis.

EIT is a general macroscopic theory based on rather broad hypotheses providing a satisfactory description of a wide variety of systems out of equilibrium. Of course, as in any macroscopic development, the equations underlying EIT contain a number of undetermined coefficients whose expression cannot be obtained from the theory itself but must be derived either experimentally or computed by means of other formalisms, like kinetic theory or statistical mechanics.

Nevertheless there remains appreciable advantages associated with the use of EIT. First, it may be compared with kinetic theory of gases in a much closer way than the local-equilibrium theory. Indeed, whereas local-equilibrium thermodynamics (or CIT) is only compatible with a first-order expansion of the distribution function of the kinetic theory, EIT is compatible with it up to the second order (and even up to higher orders if one includes in subset F higher-order fluxes).

Second, replacing the Fourier-Stokes-Newton laws by time-evolution equations for the fluxes is similar to the point of view of modern statistical mechanics. Indeed, fast phenomena are currently described by constitutive equations expressed in terms of memory functions, such as

$$q(t) = -\int_{-\infty}^{t} \lambda(t - t')\nabla T(t')\,dt' , \tag{10}$$

where $\lambda(t - t')$ is the memory function. According to Eq. (10), the value of the heat flux at time t does not only depend on the value of ∇T at time t but also on the whole history of ∇T from $t = -\infty$. It can easily be checked that Eqs. (6) and (10) are equivalent under the conditions which assure that the memory function is exponential and given by:

$$\lambda(t - t') = -(\lambda/\tau)\exp\left[-(t - t')/\tau\right].$$ (11)

Moreover, a fundamental result in linear response theory is that the memory functions are related to the time-correlation function of the fluctuations δq of the fluxes around equilibrium by

$$\lambda(t - t') = (k_B T^2)^{-1}\langle\delta q(t)\delta q(t')\rangle_{eq} = (k_B T^2)^{-1}\langle\delta q(0)\delta q(t - t')\rangle_{eq},$$ (12)

where $\langle...\rangle_{eq}$ means equilibrium average and k_B is the Boltzmann's constant. In principle, the evolution of $\delta q(t)$ can be determined by the Liouville equation but, given the complexity of the problem, it is usual in practical problems to model phenomenologically the evolution of δq. Such a modelling is precisely one of the main aims of the analysis of the evolution equations for the fluxes in EIT.

Finally, it is worth noticing that EIT is not only useful for studying fast phenomena; it provides also interesting and original information about the nature and properties of nonequilibrium steady states. EIT is a discipline which establishes a connection between dynamics and thermodynamics. Starting from the dynamics of the fluxes, i.e. the evolution equations, one can determine an expression for the nonequilibrium entropy, which depends itself on the fluxes. Therefore, in a nonequilibrium steady state, both the entropy and its resulting equations of state will depend on the steady state values of the fluxes, in contrast with the classical theory, where as a consequence of the local-equilibrium hypothesis the thermodynamic equations of state are the same as in equilibrium.

3. Illustration: Heat Conduction in Rigid Solids

To provide an explicit illustration of EIT, let us consider the simple problem of unsteady heat conduction in an isotropic non-deformable solid.

3.1. THE STATE VARIABLES

According to the hypotheses of EIT, heat conduction is locally defined by the following two variables: specific internal energy u (or temperature T) and heat flux vector q. In some occasions it may be necessary to add also higher-order fluxes as independent variables.

3.2. THE EVOLUTION EQUATIONS

The evolution equation for u is the classical energy balance: in absence of energy supply, it is given by

$$\rho\frac{\partial u}{\partial t} = -\frac{\partial q_i}{\partial x_i}. \tag{13}$$

To obtain an evolution equation for the heat flux components q_i, we assume that it obeys a first-order time evolution equation, of the form

$$\rho\dot{q}_i = -\frac{\partial J_{ij}^q}{\partial x_j} + \sigma_i^q. \tag{14}$$

J_{ij}^q is the flux of q_i, and σ_i^q the corresponding source term. In a linear theory, the most general expressions for J_{ij}^q and q_i are:

$$J_{ij}^q(u, q_i) = a(u)\delta_{ij}, \tag{15}$$

$$\sigma_i^q(u, q_i) = b(u)q_i, \tag{16}$$

where $a(u)$ and $b(u)$ are undetermined functions of u. Substituting Eqs. (15) and (16) in Eq. (14) and setting

$$b = -\frac{\rho}{\tau}, \qquad \frac{\partial a}{\partial u} = \frac{\rho\lambda}{c_v\tau},$$

with c_v the heat capacity per unit mass, one recovers the familiar Maxwell-Cattaneo equation expressing the time evolution of the heat flux vector, namely

$$\tau\dot{q}_i = -\lambda\frac{\partial T}{\partial x_i} - q_i, \tag{17}$$

which reduces to the classical Fourier law by setting the relaxation time τ equal to zero; clearly λ can be identified with the heat conductivity.

3.3. RESTRICTIONS PLACED BY THE SECOND LAW AND THE CONVEXITY REQUIRE-MENT

Restrictions on the sign of the thermophysical coefficients are provided by the second law of thermodynamics expressing that the entropy production σ^s is a non-negative quantity:

$$\sigma^s \equiv \rho\dot{s} + \frac{\partial J_i^s}{\partial x_i} \geq 0, \tag{18}$$

in which s and J_i^s (the entropy flux) are unknown functions of both u and q.

The most general (linear) expression for J_i^s is the same as in the classical theory

$$J_i^s = \frac{1}{T}q_i, \tag{19}$$

while s and σ^s take the form

$$s(u,q) = s_{eq}(u) - \frac{\tau}{2\rho\lambda T^2}q \cdot q, \tag{20}$$

$$\sigma^s = \frac{1}{\lambda T^2}q \cdot q \geq 0, \tag{21}$$

where s_{eq} is the local equilibrium value of s, depending only on u; since σ^s is positive definite it is inferred that $\lambda > 0$. Note that expressions (19-21) satisfy (18) up to the second order in the fluxes. A detailed derivation of Eq. (20) can be found in references [4, 8].

From the convexity property of s at equilibrium, which amounts to the requirement that

$$\left(\frac{\partial^2 s}{\partial q_i \partial q_i} \right)_{eq} < 0,$$

it is found that $\tau > 0$ which, combined with $\lambda > 0$, guarantees that the evolution equation (17) for q will result in a hyperbolic differential equation for the temperature: the latter is given by

$$\tau\frac{\partial^2 T}{\partial t^2} + \frac{\partial T}{\partial t} = \frac{\lambda}{\rho c_v}\frac{\partial^2 T}{\partial x_i^2} \tag{22}$$

and easily obtained by eliminating q_i between Eqs. (13) and (17). Equation (22) yields for the speed U of thermal pulses $U = (\lambda/\rho c_v \tau)^{1/2}$. It is seen that when τ vanishes the speed of thermal pulses diverges. It should be noted that another way to obtain finite speeds of propagation is to assume parabolic nonlinear equations. However, this strategy is more restricted than the one using hyperbolic equations.

The above results deserve some comments:

i. The entropy expression (20) is not the local-equilibrium entropy given by $s_{eq}(u)$ but it contains a supplementary contribution in $q.q$. This nonequilibrium contribution is explicitly identified in terms of the physical parameters τ and λ. When the relaxation time τ vanishes, Eq. (8) reduces to the local-equilibrium entropy. Thus the presence of τ establishes the connection between dynamics (evolution equation of flux) and thermodynamics (expression of entropy).

ii. The entropy flux is here equal to its classical expression $T^{-1}q$, but it would contain supplementary terms if one examines the coupling between thermal and viscous effects or in a non-local description or if one takes into account third-order terms in q.

iii. Though the set of equations (6)-(7) is very simple, it leads to a much more complicated dynamics if it is assumed that q is composed of a sum of terms q_α, each q_α obeying an evolution equation of the form of (6) and contributing to the expressions for the entropy and the entropy flux. Now the new relaxation time τ_α as well as the transport coefficients λ_α do not need to be equal to each other, but they may obey, for instance, some scaling law as $\tau_\alpha \approx \tau_0 a^\alpha$, with τ_0 and a constants. In this case, the total flux q no longer relaxes as a simple exponential but as a stretched exponential, i.e. $\exp[-(t/\tau)^b]$ with $0 < b < 1$.

4. EIT: Some Open Problems

Each new theory opens a new perspective into a set of problems. Here, we will mention some of the open problems that are conceptually or practically important for the development of the theory in the next few years. These refer to: 1) the physical meaning of the nonequilibrium equations of state; 2) the nonlinear generalization of the theory for high values of the fluxes; 3) the connection between thermodynamic stability and the positiveness of temperature and concentration profiles; 4) the role of higher-order fluxes in the description of nonequilibrium states; 5) phase transitions in steady non-equilibrium states; and 6) hydrodynamic description of nanometric electronic devices.

4.1. PHYSICAL MEANING OF THE NONEQUILIBRIUM EQUATIONS OF STATE

The equations of state are obtained as first-order derivatives of the fundamental thermodynamic potential with respect to its natural variables. For instance, in equilibrium thermodynamics, the temperature, pressure and chemical potential are related to the partial derivatives of the entropy $S(U,V,N_i)$ by

$$T^{-1} = (\partial S/\partial U)_{V,N_i}, \qquad pT^{-1} = (\partial S/\partial V)_{U,N_i}, \qquad \mu_i T^{-1} = -(\partial S/\partial N_i)_{U,V,N_j}. \qquad (23)$$

For ideal, nondegenerate, monatomic gases, this definition of absolute temperature is consistent with its identification in the kinetic theory of gases through the well-known relation

$$\left\langle \tfrac{1}{2} mc^2 \right\rangle_{eq} = \tfrac{3}{2} k_B T . \tag{24}$$

Relation (24) is used in nonequilibrium as a definition of absolute temperature. However, if one generalizes the entropy by including the fluxes, one has for the corresponding partial derivative of the generalized entropy with respect to the internal energy

$$\theta^{-1} = (\partial s/\partial u)_{\rho, q} = T^{-1} - (\partial \alpha/\partial u) q.q, \tag{25}$$

with $\alpha = (\tau_1/2\rho\lambda T^2)$, according to Eq. (20). A basic question is whether θ has a physical meaning or whether, on the contrary, it is only a mathematical artifact. In principle, this question is not exclusive of EIT: it arises in the context of kinetic theory and in rational thermodynamics as well. However, both theories do not include this question among their topics of interest.

In rational thermodynamics, absolute temperature is considered as a primitive quantity, which may depend on the temperature gradient itself, and which is recorded by small and fast thermometers. Unfortunately, not much attention has been paid to the analysis of the differences between this temperature and the local-equilibrium temperature. On the other hand, from the point of view of kinetic theory, one usually focuses more effort on the derivation of transport coefficients than to the entropy function consistent with such equations. Since the temperature has been defined from the start as the local-equilibrium temperature, no effort is made to analyse the derivatives of the entropy. The point of view adopted in the kinetic theory of gases is certainly consistent. The only missing point is the question of whether a thermometer in a nonequilibrium steady state will really measure the local-equilibrium temperature. This is always assumed to be the case as a matter of course.

However, it follows from EIT that the heat flux in the steady state is related to $\nabla\theta$ rather than to ∇T [13]. Since the difference between both terms is of higher-order in the deviations from equilibrium, the difference between θ and T is not discussed in the literature but in fact it cannot be avoided when trying to generalize the kinetic theory beyond the first-order regime. The fact that q depends on $\nabla\theta$ rather than on ∇T means that if we connect two systems at a position such that both of them have locally the same T, i.e. the same mean energy, but different θ (for instance, one of them may be in internal equilibrium whereas the other one may be in a nonequilibrium steady state characterized by some nonvanishing value of the heat flux), some heat will flow from the equilibrium system to the steady-state system, because θ is lower than T in such a system. The difference between θ and T for CO_2 at 300 K and 0.1 atm under a heat flow of 10 W/cm^2 is of the order of 0.1 K [13]. However, the difficulty in experimentally testing θ is of conceptual order: one does not know in principle whether a thermometer will measure θ or T. This is precisely the question to be clarified! If the heat flux is proportional to $\nabla\theta$ rather than to ∇T, a thermometer will measure θ rather than T because the heat flux between the thermometer and the system will vanish only when they are at the same θ rather than at the same T. Molecular dynamic analyses of the situation mentioned above could provide an answer to this conceptually subtle question. Rough qualitative estimates based on elementary kinetic theory seem to confirm that it is not T which is measured by a thermometer in a nonequilibrium steady state [13]. The meaning of thermodynamic pressure and of chemical potential under nonequilibrium conditions is also interesting and will be commented later.

4.2. BEHAVIOUR AT HIGH VALUES OF THE FLUXES

Equation (20) for the entropy is restricted to the second order in the fluxes. To deal with situations with high values of the fluxes, or to analyse fluctuations around non-equilibrium steady states, a more general formulation is needed. An elegant way to obtain it relies on information-theoretical formalisms, which maximize the entropy of the system subject to several restrictions imposed on the system [4,14]. For instance, when the mean energy of the system per unit volume u and the heat flux q across the system are known, maximum entropy arguments yield the following expression for the distribution function of the system

$$f = Z^{-1} \exp[-\beta H - \gamma J]. \tag{26}$$

Here, H and J are the microscopic operators for the energy and the heat flux, and β and γ are the corresponding Lagrange multipliers which take into account the restriction that the mean values of H and J are the required values u and q. In equilibrium, the mean value of the heat flux is zero and one has only the term βH in the exponential, i.e. the well-known canonical distribution function with β related to the equilibrium absolute temperature T by $\beta = (k_B T)^{-1}$. In Eq. (26), Z is the partition function, which is defined by requiring f to be normalized to unity. The entropy may be written in terms of Z, β and γ as

$$S = k_B[\beta u + \gamma.q + \ln Z] \tag{27}$$

and its differential form is given by

$$dS = k_B \beta du + k_B \gamma.dq. \tag{28}$$

It is instructive to examine Eqs. (26)-(28) for a specific system. For instance, for an ideal monatomic ultrarelativistic gas one has for β and γ in terms of u and q

$$\beta = \frac{3}{u} \frac{1}{\sqrt{4-3x^2}-1}, \qquad \gamma = -\frac{9x}{u} \frac{1}{2-3x^2+\sqrt{4-3x^2}}. \tag{29}$$

For a harmonic chain closed as a ring one has

$$\beta = \frac{1}{u} \frac{1+x^2}{1-x^2}, \qquad \gamma = -\frac{2x}{u} \frac{1}{1-x^2}. \tag{30}$$

In both expressions, $x = q/cu$, with c the speed of light for the relativistic gas and the speed of phonons in the harmonic chain and u is the internal energy per unit volume.

Besides providing a nonlinear generalization of the theory, several important conceptual points may be emphasized:

i. The Lagrange multiplier β is no longer related to the local-equilibrium temperature T. The latter would be defined through the expressions for the internal energy $u = 3nk_BT$ (for the relativistic gas) or $u = nk_BT$ (for the one-dimensional harmonic chain). However, as seen in Eqs. (29) and (30), β is no longer given by $\beta = (k_BT)^{-1}$, but it depends also on the heat flux. If the Lagrange multiplier β is to be related with an absolute temperature, such a temperature θ will be that of EIT, and not the local-equilibrium temperature.

ii. The parameters β and γ diverge as the heat flux tends to the critical value $q = uc$. This can be easily understood. Since the particles travel at speed c, the maximum attainable value of the heat flux is the product of the energy density by the speed of light. The fact that β diverges indicates that the generalized temperature θ tends to zero. The generalized temperature may also be considered as an illustration of the so-called contact temperature [14].

iii. Whereas the microscopic theory is able to compute explicitly the values of the Lagrange multipliers, it is unable to provide a macroscopic interpretation for them. To achieve this goal, one needs to compare expression (28) with the Gibbs equation obtained from thermodynamics. For low values of the fluxes, when only second-order terms are needed, EIT provides a physical interpretation for the multipliers, namely, $\beta = (k_B\theta)^{-1}$ and $\gamma = -(\tau_1/2\lambda T^2)q$. For more general situations, the interpretation of γ may be more complicated and it should be studied in detail.

In general, the expression for f derived from information theory does not coincide with the one derived as an exact solution of kinetic equations, as the Boltzmann equation. This is not surprising, as maximum-entropy arguments do not intend to give a detailed description for all the microscopic variables in the system, but only of those variables whose macroscopic value is controlled by the observer, but it has raised some discussions as to whether the entropy of information theory satisfies a second law when the dynamics of the variables is described by the evolution equations derived from the kinetic equations. In the phenomenological formulation, we demand that the evolution equation for q is compatible with the second law of thermodynamics as described through a positive definite production of the generalized entropy.

4.3. THERMODYNAMIC STABILITY AND POSITIVENESS OF TEMPERATURE PROFILES

The Maxwell-Cattaneo equation leads to the telegrapher's equation, which, as it is well known, does not guarantee that the temperature profile or the concentration profile will remain everywhere positive along the evolution of the system. This has been seen sometimes as an argument against the use of the Maxwell-Cattaneo equation. Recently, however, an interesting connection between the conditions of thermodynamic stability and the conditions for the positiveness of solutions of the telegrapher's equation, which had not been previously taken into account up to our knowledge [15], has been pointed out. Indeed, when one starts from a generalized entropy of the form

$$s = s_{eq} - (\alpha/2)q \cdot q, \qquad (31)$$

and one evaluates the second differential $\delta^2 s$ in a steady state under a heat flux q_0, one finds that

$$\delta^2 s = -\left[\frac{1}{cT^2} + \frac{1}{2}q_0^2 \frac{\partial^2 \alpha}{\partial u^2}\right](\delta u)^2 - 2q_0 \frac{\partial \alpha}{\partial u}\delta u.\delta q - \alpha \delta q.\delta q \tag{32}$$

is definite negative only for values of q less than a critical value q_0. For constant τ, ρ and λ, this critical value for the magnitude of the heat flux is given by $q_c = \rho u U$, with u the internal energy per unit mass and U the speed of heat pulses.

If one starts with a temperature profile everywhere positive, the profile will always remain everywhere positive in the diffusion equation, but not in the telegrapher's equation, where it may become negative at some intervals. The interesting point is that if one imposes the condition that the initial profile must satisfy not only the positiveness of T but also the restriction that the initial profile for the heat flux must be everywhere less than the critical value q_c, then the temperature profile remains everywhere positive along the evolution of the system if this obeys the telegrapher's equation. To our knowledge, this result has been shown not in general terms, but only for a special cases. It would be interesting to know whether this result is completely general, and how it should be extended to situations where τ, ρ or λ are no longer constant parameters, but depend on T.

4.4. HIGHER-ORDER FLUXES

A basic question in the description of nonequilibrium systems concerns which variables must be used. Here we focus our attention on a rigid heat conductor. In the simplest version, the evolution equation of q could be given by the Maxwell-Cattaneo equation (6). However, this equation does not have the general form of an evolution equation, which, in general terms, reads as in (4). However, in the Maxwell-Cattaneo equation the term in the divergence of the flux is lacking. Therefore, one should in principle introduce a term of this form and write

$$\rho \dot{q} = -(\rho/\tau_1)(q + \lambda \nabla T) - \nabla.Q , \tag{33}$$

with Q the flux of the heat flux. From an experimental point of view, Q is never directly measured, but from the conceptual side its meaning is clear, as it describes spatially non-local effects in the evolution equation for q. In fact, introduction of Q is essential for the description of heat transport in the regime of the so-called phonon hydrodynamics. Comparison with kinetic theory shows that Q is related to the fourth moment of the nonequilibrium distribution function.

Thus, the question arises as to whether u and q suffice for the local description of the nonequilibrium state, or whether u, q and Q are also needed, or whether u, q, Q, the flux of Q (which we will denote as $Q^{(3)}$) and so on should be included. The same question arises in kinetic theory of gases, as for example in Grad's procedure. To deal with this situation, one may propose in EIT a description based on u, and $Q^{(n)}$, with $Q^{(1)} = q$, and $Q^{(2)} = Q$ and so on, $Q^{(n)}$ being a tensor of order n ($n = 1$ to ∞). We may write for the entropy and the entropy flux

$$ds = \theta^{-1}du - \sum_n \alpha_n Q^{(n)} : dQ^{(n)} , \qquad (34)$$

$$J^s = \theta^{-1}Q^{(1)} + \sum_n \beta_n Q^{(n+1)} : Q^{(n)} , \qquad (35)$$

in which the colon (:) stands for the total contraction of the corresponding tensors. The evolution equations of $Q^{(n)}$ may be found to be, in the linear approximation, a hierarchy of the form

$$\rho\dot{Q}^{(n)} = -(\rho/\tau_n)\left[Q^{(n)} + (\beta_{n-1}\tau_n/\alpha_n)\nabla Q^{(n-1)}\right] + (\rho\beta_n/\alpha_n)\nabla . Q^{(n+1)} \qquad (36)$$

The basic problem now is to cut this hierarchy in order to have a closed set of equations for the system. The procedure will depend on the spectrum of relaxation times τ_n. If there is a set of $Q^{(n)}$ whose relaxation time is much longer than for $Q^{(n+1)}$, one may cut the hierarchy at the nth level, provided one is working at frequencies much lower than τ_{n+1}^{-1}. This is the situation found in solids at low temperature, where a description based on u, q and Q is satisfactory for the description of thermal waves and phonon hydrodynamics. However, in kinetic theory of ideal gases, all the relaxation times τ_n are of the same order of magnitude, so that in principle one should include into the formulation an infinite number of moments. Instead of truncating the hierarchy by assuming only a finite number of fluxes, we have proposed to assume, as a simple model, that all the relaxation times are equal. In this situation, and using asymptotic expressions for a continued-fraction expansion of the frequency and wavelength-dependent thermal conductivity, one may obtain in the high-frequency limit the following speed of the heat pulses [16]

$$U^2 = (\lambda/\rho c_v l^2)\left[(\lambda\tau_1/\rho c_v l^2) - 1\right]^{-1} . \qquad (37)$$

Comparison of Eq. (37) with the expression for the speed of thermal waves obtained from the Maxwell-Cattaneo equation (6), namely $U^2 = \lambda/\rho c_v \tau_1$, one may define an effective relaxation time τ_{eff} by equating (37) to $U^2 = \lambda/\rho c_v \tau_{eff}$. For monatomic gases this yields $\tau_{eff} = (11/20)\tau_1$, with τ_1 the relaxation time for q found in Grad's thirteen moment expansion. This effective value leads to a more satisfactory agreement with experimental data than the results obtained directly from Grad's method.

Thus, though in principle an infinite number of higher-order fluxes or of higher-order moments of the distribution function would be needed to describe the nonequilibrium steady state, from the practical point of view it will suffice to introduce a few fast variables as representatives of the whole set of fast variables, provided one suitably renormalizes the relaxation times of the remaining variables.

4.5. PHASE TRANSITIONS IN STEADY NONEQUILIBRIUM STATES

We have already mentioned some theoretical aspects of nonequilibrium equations of state for temperature. More interesting from the practical point of view is the non-equilibrium chemical potential which appears, for instance, in the analysis of fluid systems under shear. The phase diagram of polymer solutions is found to present a shift in the demixing critical temperature of the order of 15 K at shear pressures of 400 Nm^{-2}. Since in many practical situations in engineering processing or moulding the fluids are not in a quiescent state but submitted to shear and elongational stresses, it is important to know which may be the effects of the nonequilibrium hydrodynamic situation on the thermodynamic stability of the system. The change in the phase diagram may be qualitatively described by using the generalized chemical potential depending on the shear pressure instead of the local-equilibrium chemical potential [17] and by using the stability condition

$$\frac{\partial \mu_i(T, p, n_j, \mathbf{P}^v)}{\partial n_i} \geq 0 . \tag{38}$$

Since μ_i may contain the viscous pressure as variable, the domain of stability may be changed in a natural way by the applied shear viscous pressure.

Furthermore, since the chemical potential of the polymer with different degrees of polymerization may be changed in different ways, this may also change the weight-distribution function of the polymer, leading to polymer degradation [18]. Another starting point to describe this process is to use rate equations, where the rate constants depend on the shear flow or on the elongational rate. Of course, such an analysis is more detailed than the analysis based on thermodynamic arguments. Whereas the direct and the inverse reaction rates may be affected independently by the flux, it would be of interest to analyse in several models how the ratio between the rate constants which gives the "equilibrium constant" of the reaction, is compatible with the predictions of extended thermodynamics.

Another situation where the analysis of nonequilibrium equations of state has much interest is nuclear physics, especially the situations concerning high-energy collisions between nuclei. Such collisions are receiving much attention with the aim of obtaining a quark deconfinement transition, leading from nuclear matter to a quark-gluon plasma. Simple hydrodynamic models based on local-equilibrium cannot be used in these situations, because the total duration of the collision between nuclei is of the order of five or ten times the mean collision time between nucleons inside the nuclei. Therefore, and if one wants to keep the simplicity of a hydrodynamic description without losing crucial nonequilibrium information, one should use causal equations of transport, with a non-vanishing relaxation time. In fact, the inclusion of nonequilibrium effects in the equation of state, which has been proposed by several authors [19,20] may lead to a change of the phase diagram of nuclear matter, which should, in principle, be taken into account in the analyses of such collisions.

4.6. HYDRODYNAMIC DESCRIPTION OF NANOMETRIC ELECTRONIC DEVICES

The description of submicronic devices requires the use of generalized transport equations for electric current and heat flux. Indeed, a description based on the Boltzmann equation for electrons in semiconductors turns out to be too complicated, so that a hydrodynamic description, i.e. a study in terms of a few number of moments of the distribution function, is compelling in practice and is in fact widely used [21,22]. This raises the problem of how to close the ensuing hierarchy for the equations for moments and which transport equations should be used.

Such equations cannot be the usual Ohm and Fourier laws since the dimensions of the device become comparable to the mean-free path of the particles, and the propagation of charges is ballistic rather than diffusive for an interesting domain of temperatures. On the other hand, since the dimensions of the system are very small, it follows that very tiny differences in electrical potential or absolute temperature will result in very high gradients of these quantities, thus making unavoidable a nonlinear analysis of transport. Amongst the relevant nonlinear features, it has been seen that flux-limited transport equations are necessary if the hydrodynamic approach is to be consistent with the results derived from Monte Carlo simulations. Extended irreversible thermodynamics has been used by Anile et al to deal with the closure problem and flux limiters [23,24]. In their approach, they have chosen as independent variables the density, mean velocity, temperature, stress tensor and heat flux, as in Grad's description of monatomic gases, and have obtained expressions for the fluxes and the collision terms which successfully generalize the expressions which are common in the usual drift-diffusion models.

5. Concluding Remarks

There are several points whose analysis could be fruitful for the development of EIT. Up to now, the formulations of the theory have been mainly perturbative, i.e. limited to second-order corrections to the local-equilibrium formulation. It would be desirable to have a more general approach analysing the problem ab initio. It is possible that a geometrical formulation based on bundle spaces could be useful for that purpose. The bundles would be defined, for instance, as the collection of nonequilibrium states which decay to a given equilibrium state after being suddenly isolated. Whereas the formulation of the second law in the subspace of equilibrium states (the base space) is well-known, the formulation of bundle spaces could probably be used to extend in a natural way the second law to the general space of nonequilibrium states of the system [25].

Another topic of much interest in the abstract setting of the theory refers to variational principles [7], which could allow Hamiltonian techniques to be applied to the formulation of nonequilibrium thermodynamics. These techniques are useful, for instance, in providing through the Jacobi identity for Poisson brackets relations among nonlinear coefficients which cannot be obtained by usual thermodynamic arguments [25].

A basic open question refers to the choice of variables. We have already commented this topic in connection with higher-order fluxes and their possible elimination through a renormalization of the effective relaxation times of the first-order fluxes. However, one should also consider, at least in principle, mode couplings amongst the different variables in the theory because, in general, the mean value of two microscopic operators is not equal to the product of the mean values of the variables, so that the product itself should be taken as a new variable independent of the two preceding variables. Thus, the possibility of extending to this problem the renormalization techniques, which are usual in the analysis of critical points, could be also explored.

It should also be mentioned that the use of different probability distribution functions (for instance, the actual, but unknown, distribution function f, the canonical maximum-entropy distribution function f_{maxent}) could allow one to define several different entropies. Eu [6] has defined a generalized entropy by mixing f_{maxent} and f, with the result that the entropy is not integrable. His claim that entropy is not integrable do not refer therefore to the entropy used in EIT, to which he gives other names, but it may be rather confusing to the reader which is not expert in the field and may cast unfounded doubts about the possibility to define a generalized entropy in nonequilibrium steady states.

It is also interesting to point out that the assumption that the entropy depends on the fluxes does not imply a unique way to study nonequilibrium systems using a thermodynamic approach. Some authors [2] start from a formalism, similar in many features to that used in rational thermodynamics, in which balance laws are included through Lagrange multipliers. This method yields results which are in many cases identical to those of the present paper, where as it is clearly shown we have adopted a point of view that may be regarded as a natural extension of the classical irreversible thermodynamics. Other authors have preferred as starting point for their analyses a microscopic description, some of them using the kinetic theory [6,9] and others information-theoretical methods [14,26]. This plurality of versions leads to some points of coincidence and necessarily to some disagreements. Before closing this section a few examples of discrepancy can be mentioned. Thus, the distribution function obtained from kinetic equations in general does not coincide with that used in information theory, the positive definite character of the generalized entropy production cannot be shown in a general way from microscopic bases, the interplay between statistical and dynamical aspects is not always clear and so on. Nevertheless, these drawbacks are not exclusive of extended irreversible thermodynamics, but they are rather general and it would be difficult to find any nonequilibrium formalism exempt from these problems.

Besides these basic theoretical problems, one could finally mention, as other possible challenges to EIT, the causal description of bulk viscous effects in cosmological models [16], the analysis of two-dimensional turbulence or the discussion of transport in glassy materials. Indeed, collision times of some species in the universe may become comparable to the characteristic cosmological times defined as the reciprocal of the Hubble parameter, during decoupling times. Two-dimensional turbulence is characterized by energy and enstrophy, and has been described in many occasions through a generalized entropy which depends on both quantities [27]. This entropy is very similar to that obtained in EIT when the antisymmetric part of the pressure tensor is included as an independent variable. Finally, since relaxation times in glasses are very long, heat or electric transport or diffusion in them could be tackled in a more direct way from EIT, which takes account of nonvanishing relaxation times, than from classical nonequilibrium thermodynamics, which assumes short relaxation times.

In summary, EIT has shown its importance providing a common ground for the extension and comparison of the previously existing nonequilibrium thermodynamic theories, and in that it can be applied to a collection of stimulating problems of a wide practical and theoretical extent.

Acknowledgements. This joint work has been partially sponsored by the Commission of the European Communities-Human Capital and Mobility Program, Contract ERB 4050 PL 920346. Two of us (D. J. and J. C-V) wish to thank the DGICyT of the Spanish Ministry of Education and Science for its financial support (PB90-0676 and PB94-0718). G. L. acknowledges the Belgian Programme on Interuniversity Poles of Attraction (PAI Nos 21 and 29) initiated by the Belgian State, Prime Minister's Office, Science Policy Programming.

References

1. Casas-Vázquez, J., Jou, D., and Lebon, G. (eds.) (1984) *Recent Developments in Nonequilibrium Thermodynamics*, Lecture Notes in Physics, Springer, Berlin.
2. Müller, I. and Ruggeri, T. (1987) *Kinetic Theory and Extended Thermodynamics*, Pitagora Editrice, Bologna,.
3. Salamon, P. and Sieniutycz, S. (eds.), (1992) *Extended Thermodynamic Systems*, Taylor and Francis, New York,.
4. Jou, D., Casas-Vázquez, J., and Lebon, G. (1993) *Extended Irreversible Thermodynamics*, Springer, Berlin.
5. Müller , I. and Ruggeri, T. (1993) *Extended Thermodynamics*, Springer, Berlin.
6. Eu, B. C. (1992) *Kinetic theory and nonequilibrium thermodynamics*, Wiley, New York.
7. Sieniutycz, S. (1994) *Conservation laws in variational thermohydrodynamics*, Kluwer Academic Publishers, Boston.
8. Jou, D., Casas-Vázquez, J., and Lebon, G. (1988) Rep. Prog. Phys. **51**, 1104-1179.
9. García-Colín, L. S. (1988) Rev. Mex. Física **34**, 344-366.
10. García-Colín, L. S. and Uribe, F. J. (1991) J. Non-Equilib. Thermodyn. **16**, 89-128.
11. Lebon, G., Jou, D., and Casas-Vázquez, J. (1992) Contemp. Phys. **33**, 41-51.
12. Jou, D., Casas-Vázquez, J., and Lebon, G. (1992) J. Non-Equilib. Thermodyn. **17**, 383-396.
13. Casas-Vázquez, J. and Jou, D. (1994) Phys. Rev. E **49**, 1040-1048.
14. Muschik, W. (1992) in W. Ebeling and W. Muschik (eds.), *Thermodynamics and Statistical Mechanics of Nonequilibrium Nonlinear Systems* World Scientific, Singapore.
15. Criado-Sancho, M. and Llebot, J. E. (1993) Phys. Lett. A **177**, 323-326.
16. Jou, D. and Pavón, D. (1991) Phys. Rev. A **44**, 6496-6499.
17. Criado-Sancho, M., Jou, D., and Casas-Vázquez, J. (1991) Macromolecules **24**, 2834-2840.
18. Criado-Sancho, M., Jou, D., and Casas-Vázquez, J. (1994) J. Non-Equilib. Thermodyn. **19**, 1-16.
19. Stöcker, H. and Greiner, W. (1986) Phys. Rep. **137**, 277-392.
20. Barz, H. W., Kämpfer, B., Lukács, B., Martinas, K., and Wolf, G. (1987) Phys. Lett. B **194**, 15-19.
21. Baccarani, G. and Wordeman, M. R. (1982) Solid-state Electronics **29**, 970-977.
22. Haensch, W. (1991) *The drift-diffusion equation and its application in MOSFET modeling*, Springer, Wien.
23. Anile, A. M. and Pennisi, S. (1992) Phys. Rev. B **46**, 13186-13193.
24. Anile, A. M. and Pennisi, S. (1992) Continuum Mech. Thermodyn. **4**, 187-197.
25. Grmela, M. and Jou, D. (1993) J. Math. Phys. **34**, 2290-2316.
26. García-Colín, L. S., Vasconcellos, A. R., and Luzzi, R. (1994) J. Non-Equilib. Thermodyn. **19**, 24-46.
27. Kraichnan, R. H. and Montgomery, D. (1980) Rep. Prog. Phys. **43**, 547-619.

EXTENDED THERMODYNAMICS AND TAYLOR DISPERSION

J. CAMACHO

Departament de Física (Física Estadística),
Universitat Autònoma de Barcelona,
08193 Bellaterra, Catalonia, Spain

Abstract. This work spands in the connection between Taylor dispersion and the thermodynamics of irreversible processes by proving that the Taylor dispersion flux is a dissipative flux of extended irreversible thermodynamics. We also show a purely coarse- grained model for Taylor dispersion based on extended irreversible thermodynamics which captures the main features of Taylor dispersion along all the time scale: asymptotic diffusive behaviour, transient anisotropy, incorporation of transverse initial conditions, and transition to irreversibility.

1. Introduction

Taylor dispersion originally arose in the study of the longitudinal dispersion of a solute suspended in a solvent which flows along a rectilinear duct [1, 2]. In 1953, Taylor found that the dispersion of solute along the tube axis for long times, i.e. for times longer than the diffusive time to explore the tube section, is simply described as a translation with the mean solvent velocity, u, plus a diffusive dispersion characterized by an effective diffusion coefficient, D_T, of the form $u^2 d^2 / D_m$, d being the section size and D_m the molecular diffusivity. The dependence of D_T as the inverse power of D_m makes Taylor dispersion a useful technique for the experimental determination of molecular diffusivities [3]. But what is really striking is the fundamental result that the combined action of a unidirectional velocity field and transverse molecular diffusion leads to a longitudinal diffusion for asymptotic long times.

Since the original papers by Taylor, much work has been done trying to generalize his results. These efforts have been addressed basically in two

55

J. S. Shiner (ed.), Entropy and Entropy Generation, 55–74.

directions. By one hand, it has been generalized in order to encompass a wide variety of physical situations, from contaminant dispersion in rivers [4], or dispersion of polymers in solution [6] to sedimentation of nonspherical particles [7]; an outstanding contribution is that of Brenner and coworkers in the so-called Generalized Taylor Dispersion [5-10], that provides a unifying macroscale description of the corresponding microscale transport processes, always in the asymptotic limit of long times. On the other hand, many researchers have seeked a description of the dispersion phenomenon at shorter times; the results in this case are not conclusive, since the infinite number of time scales involved in the problem makes it very cumbersome, and the description of the dispersion at the macroscale level often becomes as complicated as in the microscale, and sometimes requires practically the same calculations [11-15].

Taylor dispersion has been faced from very different points of view: as a hydrodynamic problem [8-14], in relation with the elimination of fast modes [15], by using projection operator techniques [16], or in the context of the theory of stochastic processes [17]. There exists, furthermore, a strong connection between Taylor dispersion and kinetic theory: the problem of finding a closed equation for the averaged concentration starting from the (more detailed) hydrodynamic equations, is analogous to the derivation of the latter from kinetic theory. Until recently, however, no relation with the thermodynamics of irreversible processes had been mentioned in the vaste literature of Taylor dispersion. The interest of such a link for Taylor dispersion is twofold. By one hand, it sheds a new light over the problem, showing that Taylor dispersion is endowed with a thermodynamic structure. On the other hand, the thermodynamics of irreversible processes supplies some equations for the solute dispersion, so that it may raise some interest in the study of preasymptotic Taylor dispersion.

In the present paper, we review our previous work on the connection between Taylor dispersion and Extended Irreversible Thermodynamics (EIT), a theory that has proved succesful in fields such as, for instance, ideal and real gases, polymer solutions or phonon hydrodynamics [18]. This link is interesting for EIT as it implies the incorporation of longitudinal dispersion phenomena to its application range, and because the time scales involved in these processes are high enough so as to put easily into evidence the consequences of extended thermodynamics. We will also see that the combination of Taylor dispersion and extended thermodynamics supplies an elegant generalization of the fundamental result by Taylor.

The plan of the paper is as follows. In section 2 we demonstrate that the Taylor flux is endowed within the structure of extended irreversible thermodynamics. In section 3, we take advantage of this fact to obtain a purely macroscale model for Taylor dispersion, which is subsequently compared to some numerical simulations; this model has the remarkable feature of capturing the essential phenomenology of Taylor dispersion along all the time scale by only including a few constant parameters of clear physical meaning. We end with some concluding remarks.

2. Thermodynamics of Taylor Dispersion

In this section we show that the Taylor flux may be viewed as a dissipative flux of extended thermodynamics, with a related entropy, entropy flux, and entropy production, the latter being positive definite along time evolution. This supposes a considerable generalization of the fundamental result by Taylor which can be stated as follows: the combination of a unidirectional velocity field and transverse molecular diffusion supplies a *dissipative flux*; in the long–time limit this flux become diffusive and one recovers Taylor's result.

The demonstration has three parts. We first find in sect. 2.1 the constitutive equations for the Taylor flux components from an analysis in the three-dimensional space. In sect. 2.2, these equations are recovered by using the formalism of EIT in a one- dimensional coordinate space [19]. In the latter section, some thermodynamic functions are assumed to exist, so that sect. 2.3 is devoted to establish them in a more solid basis by showing that they can be regarded as section averages of the corresponding tridimensional thermodynamic functions.

2.1. TAYLOR FLUX COMPONENTS

For the sake of simplicity, we restrict ourselves to flows between parallel plates separated by a distance d. This choice presents the advantage of reducing the problem to two dimensions and simplifying the mathematics. The extension to arbitrary geometries is straightforward by using the corresponding eigenfunctions.

For a dilute solution, the bidimensional concentration distribution $c(x, y, t)$ — y denoting the coordinate normal to the plates — satisfies the convection–diffusion equation

$$\frac{\partial c}{\partial t} + v(y, t)\frac{\partial c}{\partial x} = D_m \left(\frac{\partial^2 c}{\partial x^2} + \frac{\partial^2 c}{\partial y^2} \right), \tag{1}$$

with $v(y, t)$ the velocity profile as seen in an inertial frame. The second term in the left-hand side describes the mass flux of the solute due to convection, and the term in the right the mass flux due to diffusion; the first flux is called convective, and the second one is said to be dissipative, because it is related to an irreversible process and consequently to an entropy production, in contrast to the first one.

After introducing the Fourier coefficients for concentration and velocity

$$c_n(x,t) = \frac{2}{d} \int_0^d dy\, c(x,y,t) \cos(\frac{n\pi y}{d}), \qquad v_n = \frac{2}{d} \int_0^d dy\, v(y,t) \cos(\frac{n\pi y}{d}), \tag{2}$$

a direct calculation yields the equations for each mode $c_n(x,t)$.

For the section–averaged concentration $c(x,t) = c_0(x,t)/2$, one has

$$\frac{\partial c(x,t)}{\partial t} + \frac{\partial}{\partial x}\left(c(x,t)u(t) + \sum_{n=1}^{\infty} \frac{1}{2} v_n c_n(x,t) \right) = D_m \frac{\partial^2 c(x,t)}{\partial x^2}, \tag{3}$$

with $u(t) \equiv (1/d) \int_0^d dy\, v(y,t)$ the mean solvent velocity. This expression may be considered as a mass balance equation with a convective term, $c(x,t)u(t)$, a unidimensional molecular diffusion flux, $J_m = -D_m\, \partial c(x,t)/\partial x$, and a velocity–dependent spectrum

$$J_n(x,t) = \frac{1}{2} v_n(t) c_n(x,t) \qquad\qquad n = 1,2,\ldots \tag{4}$$

Therefore, J_n is the contribution of the n–th mode to a Taylor flux defined as

$$J_T(x,t) = \frac{1}{d} \int_0^d dy\, c(x,y,t) v(y,t),$$

obviously the expression that one should expect for a particle flux related to the velocity field. Our purpose is to demonstrate that the flux J_T, which has a convective character when it is regarded from three dimensions – and, consequently, there is no irreversibility associated to it –, turns to be dissipative when seen in one dimension – being then related to an irreversible process and to an entropy production –, at the same level of the longitudinal diffusion flux, but of course with several distinguishing features which are analyzed in the next subsection.

Finally, the equations for the higher-order modes can be arranged as equations for $J_n(x,t)$:

$$\dot{J}_n + \left(\tau_n^{-1} - \gamma_n\right) J_n + \frac{\partial}{\partial x} \sum_{m=1}^{\infty} \gamma_{mn} J_m = -\frac{D_n}{\tau_n} \frac{\partial c(x,t)}{\partial x} + D_m \frac{\partial^2 J_n}{\partial x^2}, \tag{5}$$

where the coefficients τ_n, γ_n, D_n and γ_{mn} are some combinations of d, D_m and v_n.

These evolution equations are much more complex than the simple Fickian laws coming from classical irreversible thermodynamics. Conversely, they lead to a nonpositive entropy production in this thermodynamic context as can be seen by introducing eq. (5) into the classical expression for the entropy production, $\sigma \propto J.\partial c(x,t)/\partial x$, with $J = J_m + J_T$ the material (or dissipative) particle flux.

2.2. ENTROPY BALANCE AND CONSTITUTIVE EQUATIONS

The purpose of the present subsection is to obtain the latter equations for J_n through a very different method. We adopt a completely one-dimensional point of view, as if we were 1D observers, and perform a unidimensional thermodynamic treatment.

In effect, the constitutive equations (5) can be obtained following the usual procedure of irreversible thermodynamics, i.e. stating an entropy balance and imposing a positive entropy production according to the local version of the second law of thermodynamics [19].

In the study of Taylor dispersion, it is generally admitted that the solution is so diluted that the barycentric velocity of the mixture coincides with the solvent speed — this property was used in writing the convection-diffusion equation (1). In this way the role of the solvent becomes essentially parametric and consists in imposing the velocity field in which the solute evolves. This allows to treat the system as monocomponent.

In extended thermodynamics, nonequilibrium states are characterized not only by the classical local–equilibrium variables (such as the specific energy or the specific volume, for an ordinary fluid) but also by the dissipative fluxes, so that the thermodynamic functions, like the entropy or the entropy flux, depend on all these quantities. Accordingly, we introduce a generalized entropy for Taylor dispersion which includes, besides the local–equilibrium magnitude – in our case, the concentration $c(x,t)$–, the Taylor contributions J_n, and their fluxes P_n, i.e. $s(x,t) = s(c,\{J_n\},\{P_n\})$. The incorporation of the second-order tensors P_n allows to introduce spatial correlations between fluxes, which is a characteristic feature in the dynamics of fluxes J_n as shown in eq.(5); variables similar to P_n have been employed in the past, for instance, in the study of phonon hydrodynamics [18] or nonfickian diffusion in polymers [24].

We thus assume for situations not far from equilibrium a generalized entropy of the form

$$s(x,t) = s_{eq}(x,t) - \sum_{n=1}^{\infty} \frac{1}{2}\alpha_n J_n \cdot J_n - \sum_{n=1}^{\infty} \frac{1}{2}\beta_n P_n : P_n, \qquad (6)$$

with $s(x,t)$ the entropy per unit volume, $s_{eq}(x,t)$ the local–equilibrium term satisfying the classical Gibbs equation at constant energy and volume,

$$ds_{eq} = -\mu T^{-1}dc, \qquad (7)$$

and the remaining terms are purely non–equilibrium contributions; the symbol : indicates tensorial contraction.

This is the simplest expression which makes use of the independent variables under consideration. The positive sign of coefficients α_n and β_n guarantees that the entropy is maximum at equilibrium. These coefficients are assumed to be independent of J_n and P_n, and their form is specific for each particular system. Although they are generally considered as independent of time, in Taylor dispersion they are expected to depend on the velocity profile so that for nonstationary flows they become time–dependent.

For the entropy flux one can distinguish three contributions. Keeping only second-order terms in the fluxes, one has

i) The classical term linked to a particle flux [21], $-\mu T^{-1}J$, with $J(x,t)$ the total dissipative mass flux, i.e. the sum of the longitudinal molecular diffusion flux, $J_m(x,t)$, and the velocity-dependent Taylor flux, J_T.

ii) A term which accounts for the anisotropy that the velocity field introduces in the dispersion of the solute. This term is assumed to couple each flux with the others: $-\sum_{n,m=1}^{\infty} \vec{\alpha}_{mn} \cdot (J_m J_n)/2$; the $\vec{\alpha}_{mn}$ are vectorial coefficients which point in the flow direction; they must necessarily be odd in the velocities in order to display this kind of anisotropy, since then the sign of this contribution becomes sensitive to a inversion of the velocity field $(v \to -v)$.

iii) And finally, the contribution of the fluxes P_n. The simplest case is to consider a term like $-\sum_{n=1}^{\infty} \delta_n P_n \cdot J_n.$, with δ_n some coefficients independent of J_n and P_n.

Other terms, like ϕJ_n, $\phi_{mn} \cdot J_m$ or $P_n \cdot \vec{\phi}_n$ — ϕ's being (velocity-dependent) coefficients of suitable tensorial order —, although mathematically allowed, are forbidden by thermodynamics because they do not lead to positive definite entropy production as can be easily proved.

The total entropy flux in a frame which moves with the one–dimensional fluid particles is thus

$$J^s(x,t) = -\mu T^{-1}J - \frac{1}{2}\sum_{n,m=1}^{\infty} \vec{\alpha}_{mn} \cdot (J_m J_n) - \sum_{n=1}^{\infty} \delta_n P_n \cdot J_n \qquad (8)$$

In a general frame will also appear the convective term $s(x,t)u(t)$. The second term in the r.h.s, which was not used in extended thermodynamics up to now, plays an important role in Taylor dispersion.

After introducing (6), (8) and the mass conservation equation

$$\frac{dc(x,t)}{dt} + \nabla.J(x,t) = 0 \qquad (9)$$

into the entropy balance equation

$$\frac{ds}{dt} + \nabla.J^s = \sigma_s \tag{10}$$

one finds for the entropy production, σ_s

$$\sigma_s = - J_m \cdot [T^{-1}\nabla\mu] -$$

$$- \sum_{n=1}^{\infty} J_n \cdot \left[T^{-1}\nabla\mu + \alpha_n \dot{J}_n + \frac{1}{2}\dot{\alpha}_n J_n + \delta_n \nabla.P_n + \sum_{m=1}^{\infty} \vec{\alpha}_{mn} \nabla.J_m \right] -$$

$$- \sum_{n=1}^{\infty} P_n : \left[\beta_n \dot{P}_n + \frac{1}{2}\dot{\beta}_n P_n + \delta_n \nabla J_n \right] \tag{11}$$

The second law requirement of positiveness of the entropy production obliges the terms inside brackets (thermodynamic forces) to be linked to the dissipative fluxes which preced them, namely J_m, J_n and P_n. The simplest relation is to consider each force to be proportional to the corresponding conjugate flux; let us denote with K_m, K_{1n} and K_{2n} the respective constants of proportionality; of course, they must be positive for the entropy production to be positive definite.

If one assumes the β_n's to vanish, some direct calculations give:

$$J_m = -D_m \nabla c,$$

$$P_n = -D_m \nabla J_n,$$

$$\dot{J}_n + \left(\tau_n^{-1} + \frac{\dot{\alpha}_n}{\alpha_n} \right) J_n + \sum_{m=1}^{\infty} \frac{\alpha_{mn}}{\alpha_n} \nabla.J_m = -\frac{D_n}{\tau_n}\nabla c + \frac{l_n^2}{\tau_n}\nabla^2 J_n, \tag{12}$$

with the identifications

$$D_n = \frac{1}{K_{1n}T}\frac{\partial\mu}{\partial c}\bigg|_T , \qquad \tau_n = \alpha_n D_n T \frac{\partial\mu}{\partial c}\bigg|_T^{-1}, \qquad l_n^2 = \frac{\delta_n^2}{K_{1n}K_{2n}}, \tag{13}$$

$$\gamma_n = -\frac{\dot{\alpha}_n}{\alpha_n}, \qquad \gamma_{mn} = \frac{\alpha_{mn}}{\alpha_n}. \tag{14}$$

Eq. (12) has the same form as eq. (5). At the sight of eqs. (12)-(14), one concludes that extended thermodynamics provides a connection between each term in the evolution equation (5) and some thermodynamic functions: the diffusive term is linked as usual to entropy production, the relaxational one is related to the nonequilibrium part of a generalized entropy, the divergence term is connected to an anisotropic entropy flux, and the term introducing the spatial correlation in fluxes is related to both an entropy flux and an entropy production.

Therefore, we have proved our previous statement that the Taylor dispersion flux is a dissipative flux. This proof is completed in the next section.

2.3. THERMODYNAMIC FUNCTIONS FOR TAYLOR DISPERSION

In the previous section, the one-dimensional thermodynamic functions (6), (8) and (11), were used to obtain the constitutive equations. Our present aim is to find the connection between these one-dimensional functions and the corresponding three- dimensional ones. To accomplish this, we procede analogously as in the derivation of macroscopic thermodynamic functions from the kinetic theory of gases; in the latter case, the molecular degrees of freedom are averaged to keep only the macroscopic quantities, whereas in the present work the transversal degrees of freedom are averaged to keep only the longitudinal one. In this way, it will be seen that the one-dimensional functions coincide with the lower-order terms of a development in the section-averages of three-dimensional quantities. This comparison is not only analytical, but also conceptual, stressing the different interpretation of the thermodynamic functions given in a 1D description, where the transversal coordiantes are absent, and in 3D, where they are present [20]. Let us stress again that though, for simplicity, we use two dimensions instead of three, the physics attached in both cases is the same as we have seen in sect. 2.1, and quite different to the approach given in sect. 2.2.

2.3.1. *Entropy*
The evaluation of expression (6) for the entropy requires the knowledge of the chemical potential of the solute, $\mu(x, t)$, both for the determination of the local equilibrium entropy, and for the estimation of coefficients α_n, which depend on $\partial\mu/\partial c|_T$ as shown by (13). Notice that in the hydrodynamic analysis summarized in Section 2.1, the solute particles are assumed noninteracting; this becomes apparent from the fact that the molecular diffusivity D_m was taken constant, which is not true for interacting particles since there exists then some dependence of D_m on the concentration. Therefore, the chemical potential corresponding to the ideal system under study has the same form as for an ideal gas, namely

$$\mu(x, t) = kT \ln c(x, t) + f(T) \tag{15}$$

k being the Boltzmann constant, and $f(T)$ a function of the temperature without interest for the present problem, which is completely isothermal. With the help of (15) and eqs. (13)–(14), we can determine the expression of any thermodynamic function. As an example, we calculate the entropy.

According to the classical Gibbs equation (7), the integration of (15) added to expression (13) for α_n yields

$$s\left[c(x,t),\{J_n\}\right] = -k\,c(x,t)\,(\ln c(x,t) - 1) - \frac{k}{c(x,t)} \sum_{n=1}^{\infty} \frac{1}{v_n^2} J_n^2. \qquad (16)$$

Let us now analyze the problem from a tridimensional point of view. In 3D (two dimensions for flows between parallel plates), the Gibbs equation describing the system is the local-equilibrium one,

$$ds(x,y,t) = -\mu(x,y,t)T^{-1}dC(x,y,t). \qquad (17)$$

Similarly as in (15), the chemical potential $\mu(x,y,t)$ has the form of the chemical potential for an ideal gas, with the only difference that the coordinate space in the present case is wider

$$\mu(x,y,t) = kT \ln C(x,y,t). \qquad (18)$$

And after integration of (17), one finds for the entropy per unit volume at position x

$$s(x,t) = -k \int_0^d \frac{dy}{d} C(x,y,t)\,(\ln C(x,y,t) - 1). \qquad (19)$$

This is also the expression stemming from information theory [22], where it is interpreted as the ignorance that the knowledge of the function $C(x,y,t)$ leaves about the transverse distribution of the solute particles.

It is useful to introduce a perturbation $\phi(x,y,t)$ defined by

$$C \equiv c(x,t)\,(1 + \phi).$$

In the limit, $\phi \ll 1$, that is to say, when transverse inhomogeneities are small as compared to the average value, eq. (19) can be arrenged to give

$$s + k\,c(x,t)\,(\ln c(x,t) - 1) = -\frac{1}{4}\frac{k}{c(x,t)} \sum_{n=1}^{\infty} c_n^2 = -\frac{k}{c(x,t)} \sum_{n=1}^{\infty} \frac{1}{v_n^2} J_n^2, \qquad (20)$$

which is identical to (16), although the interpretation of the entropy given in each case is quite different. In the first case, the approach is 1D, and the entropy is expressed in terms of one-dimensional quantities, $c(x,t)$ and $J_n(x,t)$; while in the second tridimensional case, $s(x,t)$ is regarded either as the sum of the entropies of the 3D fluid elements or, within the context of information theory, as a measure of the order of the solute distribution across the section.

2.3.2. *Entropy Flux*

In the latter section, we used expression (8) for the entropy flux which contains two new terms compared to the classical one. The introduction of such terms proved to be useful in the obtention of the constitutive equations for $J_n(x,t)$. The aim of this section is to establish the connection between these terms and the entropy flux defined in three dimensions. This kinetic comparison bears special interest for the term related to the transient anisotropy because it is the first time that a contribution of this type is included in the formalism of extended thermodynamics.

In the tridimensional space, the x-component of the entropy flux writes as

$$J_x^s(x,y,t) = s(x,y,t)v(y) - \mu(x,y,t)T^{-1}J_{mol}^x \qquad (21)$$

When integrated over the section, we have two contributions, one linked to molecular diffusion, the other coming from convection.

Making use of $J_{mol}^x = -D_m\partial C/\partial x$, eq. (18) for $\mu(x,y,t)$ and $C = c(1+\phi)$, it can be seen that the diffusive contribution, up to second order in the fluxes, provides two terms. One is $-\mu(x,t)T^{-1}J_m(x,t)$, with $\mu(x,t) = kT\ln c(x,t)$ the local- equilibrium chemical potential, and the second one coincides with the term $P_n.J_n$.

On the other hand, the convective contribution to the entropy flux, restricting again up to second order terms in the fluxes, supplies two other terms. One has the classical form for an entropy flux associated to a mass flux J_T, i.e. $-\mu(x,t)T^{-1}J_T(x,t)$. This is quite remarkable: similarly as the mass flux in 3D has two contributions — the convective and the dissipative ones — but in 1D the first one is seen as dissipative, with the entropy flux occurs the same, the convective term conspires to give a contribution to J^s which has the form of a typical entropy flux associated to a dissipative flux.

This convective term supplies as well the one-dimensional contribution of the entropy flux associated to anisotropic dispersion, the second term in the r.h.s of (8), what seems quite natural since it is convection the responsible of the anisotropy.

We have thus shown that the same expression for the entropy flux is obtained through a classical thermodynamic formalism in three dimensions and extended thermodynamics in one dimension for situations close enough to equilibrium.

2.3.3. *Entropy Production*

Finally, the classical expressions of entropy and entropy flux in 3D, eqs. (17) and (21), obviously lead to an usual entropy production

$$\sigma_s(x, y, t) = \frac{1}{k_m T} J_{mol}^2(x, y, t),\tag{22}$$

where the dissipative coefficient k_m is linked to D_m through

$$k_m = \frac{1}{D_m T} \frac{\partial \mu(x, y, t)}{\partial C(x, y, t)}\bigg|_T = \frac{1}{D_m T} \frac{1}{C(x, y, t)}.\tag{23}$$

Therefore, one can write $\sigma_s(x, t)$ as

$$\sigma_s(x, t) = k D_m \int_0^d \frac{dy}{d} \frac{1}{C(x, y, t)} \left[\left(\frac{\partial C(x, y, t)}{\partial x} \right)^2 + \left(\frac{\partial C(x, y, t)}{\partial y} \right)^2 \right]\tag{24}$$

The first term in the right-hand side, describing the entropy production due to longitudinal diffusion, can be proved to yield the one-dimensional contributions in J_m^2 and P_n^2, the latter term expressing the correction due to the non-uniformity of the solute distribution along the section.

Finally, the last term in the integral (24) can be easily rearrenged to yield the terms in J_n^2 to the unidimensional entropy production, so that one realizes that the dissipation vinculated by the Taylor flux proceeds from transverse molecular diffusion. This is quite reasonable, since the Taylor flux J_T arises from the coupling of the velocity profile and transverse molecular diffusion.

3. A Macroscale Model for Taylor Dispersion

In Section 2, we have established a rigorous proof that the Taylor dispersion flux is a dissipative flux of extended thermodynamics. In the present one, we use this result in order to obtain a model for Taylor dispersion valid for short times, thus going further than the asymptotic analysis given by Taylor.

The study of Taylor dispersion at intermediate times has been performed in the past by other authors [11-15]. All of them start from the tridimensional convection–diffusion equation

$$\frac{\partial c(x, t)}{\partial t} + u \frac{\partial c(x, t)}{\partial x} + \frac{\partial}{\partial x} \left[\overline{\Delta c \Delta v} \right] = D_m \frac{\partial^2 c(x, t)}{\partial x^2}\tag{25}$$

where $\Delta v \equiv v(y, z) - u$ and $\Delta c \equiv C(x, y, z, t) - c(x, t)$ are the deviations from the mean (section–averaged) values of the velocity field and the concentration profile, respectively (the overbar denotes section averaging). On the other hand, they make some hypotheses on the dependence of the cross–sectional variation Δc in terms of the average concentration $c(x,t)$, following Taylor ideas.

Our purpose is to obtain the simplest constitutive equation capturing the main features of the longitudinal dispersion of the solute along all the time scale. In contrast to the previous approaches, ours is completely one-dimensional [23].

The section is organized as follows. Firstly, we derive a constitutive equation for the Taylor dispersion flux from the formalism of extended thermodynamics, and afterwards we compare the theoretical predictions of our model for the lowest moments of the concentration with numerical simulations.

3.1. A PURELY MACROSCALE MODEL FOR TAYLOR DISPERSION BASED IN EIT

As we have seen, to obtain an exact one-dimensional description of the solute dispersion in a tube, we have to deal with an infinite set of coupled constitutive equations — one for each mode — and, accordingly, an infinite number of transport coefficients. In this situation one can seriously wonder which are the advantages of working in the coarse-grained (one-dimensional) space instead of the complete three-dimensional space, where the dispersion follows a simple convection–diffusion equation. Our goal is to establish an approximate equation by compacting the infinite number of modes constituting the Taylor dispersion flux into one single mode, the Taylor flux itself $\mathbf{J}_T$, considered as a single quantity [23].

In the simplest case of Taylor dispersion under isothermal and incompressible conditions, one has the following generalized Gibbs equation

$$ds(x,t) = -\mu(x,t)T^{-1}\,dc(x,t) - \alpha\,\mathbf{J}_T\cdot d\mathbf{J}_T, \qquad (26)$$

$s(x,t)$ being the entropy per unit volume, and α a scalar coefficient which does not depend on $\mathbf{J}_T$. The first term in the right–hand–side corresponds to the local equilibrium contribution, and the second one is a flux–dependent purely nonequilibrium part of the generalized entropy.

Similarly, a generalized entropy flux is introduced as a general function of the dissipative fluxes. We add to the classical expression for the entropy flux, $-\mu(x,t)T^{-1}\mathbf{J}_T$, a term aiming to describe the transient anisotropy in the solute dispersion induced by the velocity profile (a well-known feature of Taylor dispersion [1]), and a term accounting for spatial correlations (although no experimental observation of them has been mentioned in the literature, it was shown in Sec. 2 that they stem from a theoretical analysis in three dimensions, so that they are included here for the sake of completeness)

$$\mathbf{J}^s(x,t) = -\mu(x,t)T^{-1}\mathbf{J}_T - \frac{1}{2}\alpha_0 J_T^2 \mathbf{u} - \delta \, \mathbf{P}_T \cdot \mathbf{J}_T, \tag{27}$$

where α_0 and δ are phenomenological coefficients with no dependence on $\mathbf{J}_T$, $\mathbf{u}$ is a vector in the direction of the flow and with modulus u, the mean speed of the solvent as seen from the frame where the tube is fixed, and $\mathbf{P}_T$ is a second-order tensor representing the flux of $\mathbf{J}_T$; this term allows us to incorporate spatial correlations. The second term in the r.h.s. accounts for the anisotropy induced by the flow, since the change $\mathbf{u} \to -\mathbf{u}$ modifies the sign of this contribution. Other terms, like $u^2 \mathbf{J}_T$ or $\mathbf{P}_T.\mathbf{u}$, although mathematically possible, are forbidden by thermodynamics because they lead to nonpositive values for the entropy production.

Combining eqs. (26) and (27) with the entropy balance equation leads to the following constitutive equation for J_T

$$J_T + \tau_T \dot{J}_T + \tau_T \beta u \frac{\partial J_T}{\partial x} = -D_T \frac{\partial c}{\partial x} + l_T^2 \frac{\partial^2 J_T}{\partial x^2}. \tag{28}$$

Aside from the classical Fickian contribution for the Taylor dispersion flux, eq. (28) contains a relaxational term characterized by the time parameter τ_T, a correlation length l_T, and an anisotropic term, the one in $\partial J_T/\partial x$, whose parity is obviously different from the one of the preceding terms, so that β is a numerical parameter characterizing the anisotropy.

The constitutive equation (28) is obviously a simplification since the Taylor dispersion flux is not a single quantity but the sum of an infinite number of partial fluxes. The following subsections are devoted to check its validity for the study of the longitudinal dispersion of a solute in suspension in a solvent flowing in a tube along all the time span by comparing with numerical simulations. To this end, eq. (28) must be combined with the mass balance equation

$$\frac{dc(x,t)}{dt} + \frac{\partial J(x,t)}{\partial x} = 0, \tag{29}$$

$J(x,t) = J_m + J_T$ being the total particle flux, which can be written as the sum of a molecular part, J_m, and a flow- dependent contribution, J_T. The constitutive equation for J_m is assumed to be Fickian

$$J_m = -D_m \frac{\partial c(x,t)}{\partial x} \tag{30}$$

since molecular diffusion relaxes in a time of the order of the collision time, much smaller than the time scale under study.

With the help of (28), (29) and (30), one finds

$$\tau_T \frac{\partial^2 c}{\partial t^2} + \frac{\partial c}{\partial t} + \tau_T \beta u \frac{\partial^2 c}{\partial x \partial t} - \left(\tau_T D_m + l_T^2\right) \frac{\partial^3 c}{\partial x^2 \partial t} =$$
$$D \frac{\partial^2 c}{\partial x^2} + \tau_T \beta u D_m \frac{\partial^3 c}{\partial x^3} + l_T^2 D_m \frac{\partial^4 c}{\partial x^4} \qquad (31)$$

with $D \equiv D_T + D_m$ the total diffusion coefficient. Let us comment on some properties of equation (31):

a) It is of the telegrapher's type (a more simplified version of it has been used in the study of solute dispersion in rivers by Thacker [25] and Smith [14]), with constant coefficients, so that although solving it may be extremely complicated the analysis of its moments becomes quite simple, as it leads to ordinary linear differential equations.

b) It asks for the incorporation of some details about the transverse initial distributions, since being a second-order differential equation in time its solution requires the knowledge of initial conditions on $\partial c / \partial t$. This is a basic point in the description of the solute dispersion in the short-time limit, highly dependent on the initial solute distribution along the section of the tube.

c) In the long-time limit it supplies the diffusive behaviour characterizing the Taylor-Aris' results.

d) It describes a transient anisotropic dispersion through the terms containing the parameter β: at intermediate times, the solute disperses differently in the sense of the flow and against it, but at long times these terms become negligible and the anisotropy disappears.

e) It provides a transition from a partially reversible regime at short times to a purely irreversible regime at long times (a similar transition has been observed experimentally in tracer dispersion in porous media [26]). Indeed, for $t \ll \tau_T$, the time reversal terms $\partial^2 c / \partial t^2$ and $u \partial^2 c / \partial x \partial t$ are relevant, so that some reversible behaviour in the dispersion of the solute is predicted at these time scales; in the long-time limit they become negligible and the behaviour turns out completely irreversible.

Finally, in order to compare with numerical simulations, we need explicit expressions for the macroscale parameters appearing in the model equation. An analysis in the tridimensional space allows to recover eq. (28) and to obtain these expressions. One finds: $\tau_T = D_T / \overline{v^2}$, $l_T^2 = D_m \tau_T$, and two possible expressions for the β parameter [23].

3.2. COMPARISON WITH NUMERICAL SIMULATIONS

For the sake of simplicity, the simulations have been performed for a flow between parallel plates [23]. In all of them, we have considered a plane

Poiseuille flow. Its basis is quite simple [27], we start by considering two parallel lines separated by a distance d and a large amount of points (N) uniformly distributed between them at $x = 0$; these points represent the solute particles. At every time step τ_s, each particle undergoes two motions: it travels a length $l_c(y) = v(y)\tau_s$, which corresponds to the convection under the solvent flow, and a length l_d – independent of y – in a random direction, simulating the two–dimensional Brownian motion of the solute particles. One lets the system evolve, one computes the value of $< \Delta x^2 >$ at different points in time, t, ($< \ldots >$ denotes the average over the simulation distribution) and represents the ratio $(\Delta x)^2(t)/2t \equiv \overline{D}(t)$ versus the penetration length of the solute, $L =< x(t) >$. We have introduced the function $\overline{D}(t)$ that may be called the "effective dispersion coefficient" since it has the dimensions of a diffusion coefficient and it tends asymptotically to D in the long-time limit. This simulation is referred to as a *transmission* dispersion simulation. The results for some specific values of the parameters are displayed in figs. 1 (squares).

Other simulations consist in letting the system evolve during a time span t_i, suddenly reverse the velocity field: $v(y) \rightarrow -v(y)$ and represent again $\overline{D}(t)$ as a function of $< x(t) >$. These are called *echo* dispersion simulations. In figs. 1 some inversion curves are displayed (diamonds) for different inversion lengths.

We may give a rough interpretation of the results shown in figs. 1. In the transmission simulation curve, one can distinguish three regions: two constant ones, one at short times (short penetration lengths) and another at long times, and a transition region between them. The constant behaviour in the short-time limit corresponds obviously to the domain of molecular diffusion, and the flat regime in the long-time limit to the completely developed Taylor dispersion.

Referring to the echo simulation curves, there are some points to comment on.

i) For very short inversion times the effective dispersion coefficient approaches the value D_m at a time twice the inversion time ($t = 2t_i$). This is understood in a very simple way: the fluid particles initially start moving with the velocity profile, inside each of them the diffusion process takes place but because of the short time of the "experiment", the fraction of solute particles that escape from the original fluid particles is negligible, so that when one reverses the velocity field and analyses the dispersion at $t = 2t_i$ every fluid particle is found at the same place that in $t = 0$ but with a diffusion process that has been occurring along all the time. The behaviour is thus partially reversible.

ii) In the long-time limit, in the regions where the transmission curve is flat the echo curves are straight lines. This means that the reversal of

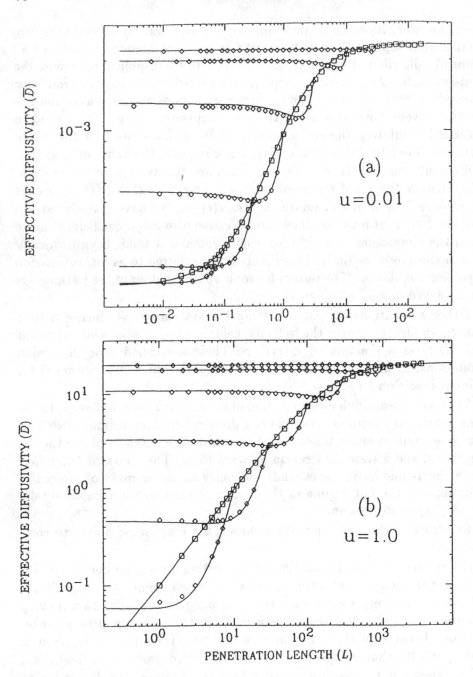

Figure 1. Comparison between theoretical predictions and simulations. Squares and diamonds correspond to transmission and echo simulations, respectively; theoretical curves are denoted by solid lines. We show the results for two different Peclet numbers ($Pe = ud/D_m$). In all the cases we have chosen $l_d = 0.03$ and $d = \tau_s = 1$; the inversion times are given in time step units: $t_i = 10, 30, 100, 300, 1000, 3000$. (a) $u = 0.01$ ($Pe = 44$), $N = 5.\,10^4$; (b) $u = 1.0$ ($Pe = 4400$), $N = 10^5$.

the velocity field does not affect the dispersion process; this is so because at long times the mixing process is so intricate, the coupling between the solute and the velocity field is so strong, that the reversal of velocities does not contribute to recover the initial distribution, but the mixing process is similar to the case occurring when the velocity field is kept unchanged. In this limit, the dispersion is thus completely irreversible.

iii) At intermediate times, in the transient region between the two pure diffusive regimes, the effect of reversing the velocity field results, essentially, in a fast decrease of $\overline{D}(t)$ followed by an increase, reaching at $t = 2t_i$ values between D_m and D which are smaller than the corresponding ones at $t = t_i$. Then the reversal of velocities leads to a stronger dispersion than the simple molecular diffusion, but smaller than in developed Taylor dispersion.

In summary, echo simulations evidenciate the existence of a transition from a partially reversible regime at short times to a completely irreversible regime at long times, in agreement with the qualitative prediction of the model equation (31).

In order to compare with the preceding numerical simulations one must calculate an expression for the effective dispersivity $\overline{D}(t)$. We can do this by using the method of moments in eq.(31) under proper initial conditions.

In figs. 1, the theoretical predictions are displayed for two different values of the mean velocity (solid lines), from values of D_T similar to D_m (fig.1a) to $D_T \gg D_m$ (fig.1b). In both cases, an excellent agreement is shown. The comparison with simulations for which initially the solute does not spread over the whole section, but only over a fraction of it, also leads to satisfactory results for spread enough initial distributions for the first and second moments [23].

4. Concluding Remarks

We have shown that a 1D observer can describe exactly the longitudinal dispersion of a solute in a tube; this is quite surprising since for such an observer no transversal notion makes any sense, so that the concepts of transverse diffusion and velocity field, which seem to be so important in the description of the solute dispersion, have no meaning for him; instead, he can use the concept of dissipative flux, as well established in one dimension as in three, and use extended thermodynamics to derive the equations for the solute dispersion, as we did in Sec. 2.2. Therefore, although both the one-dimensional and the three-dimensional observers find the same equations for the dynamics of the fluxes, the concepts and quantities used by each of them are quite different. While a 3D researcher expresses the solute dispersion as a combination of molecular diffusion and the convection of the solvent embodied in the velocity field, the 1D scientist talks in terms of dissipative fluxes, one flux related to molecular diffusion and the other related to convection; consequently, a mass flux which is seen as convective in 3D is viewed as dissipative in 1D. This dissipative flux cannot be included in the classical description of thermodynamics of irreversible processes, but is naturally incorporated in extended irreversible thermodynamics.

By analogy with the mass fluxes, which have different interpretations in one and three dimensions, the thermodynamic functions are regarded differently by 1D and 3D observers. In 1D, these functions are linked to the dynamics of the fluxes, with no reference to transversal coordinates, but in 3D, they are averages over the tube section of the corresponding tridimensional quantities. While in 1D, it is required to use extended thermodynamics in order that the entropy production be positive definite, in 3D the thermodynamic formalism is the classical one, because the molecular diffusion flux is characterized by a vanishing relaxation time at the time scales under study.

Apart from the fundamental result that the Taylor flux is a dissipative flux of extended thermodynamics, we have proposed a purely global model for Taylor dispersion. Contrary to previous approaches, where the coarse-grained equations are found from a study in the three-dimensional space, ours is completely one-dimensional. The macroscale equation thus obtained, eq.(31), is not exact but the simplest one capturing the main features of Taylor dispersion along all the time span. An analysis in three dimensions allows to recover the macroscale equation, thus getting a microscale confirmation and supplying some explicit expressions for the general transport coefficients appearing in it.

Comparison with numerical simulations has shown that our macroscale scheme works well over all times for the evolution of the mass center and the effective diffusivity for spread enough initial distributions; the fit is also good at intermediate times for any initial distribution. These points, in addition to the introduction of transverse initial conditions (which arise from the fact that our model equation is second- order in time), of the anisotropy induced by the velocity profile, the description of the transition to irreversibility, and the asymptotic diffusive behaviour, allow to conclude that our — conceptually and operatively — simple macroscale model captures the essential features of Taylor dispersion within all the time interval.

From the viewpoint of levels of description, we have presented here a macroscale framework (i.e. which works in a reduced coordinate space) that leads to a macroscale equation with general phenomenological coefficients (D, τ and β). The explicit form of these coefficients in terms of more fundamental quantities depends on the specific system under study (tube geometry, porous media,...), which implies an analysis in the complete coordinate space. This is analogous to the relation between thermodynamics and kinetic theory. It is nice to see that in both coarse–graining processes the resulting equations have essentially the same form: relaxational equations for the fluxes with Fickian behaviour at long times, and telegrapher–like equations for the specific parameter under study, with asymptotic diffusive behaviour.

Acknowledgements

The author whishes to express his gratitude to Professors D. Jou, J. Casas-Vázquez and G. Lebon for fruitful comments and discussions. This work was partially done in the Université de Liège under the support of a EC grant inserted in the E.C. program ERB CHR XCT 920007. Financial support is also acknowledged to the Spanish Ministry of Education under a FPI grant.

References

1. Taylor, G.I. (1953) *Proc. R. Soc. London*, Ser. A **219**, 186; (1954) A **223**, 446.
2. Aris, R. (1956) *Proc. R. Soc. London*, Ser. A **235**, 67.
3. Alizadeh, A., Nieto de Castro, C.A., and Wakeham, W.A. (1980) *Int. Journal Thermophys.* **1**, 243; Alizadeh, A. and Wakeham, W.A. (1982) *Int. Journal Thermophys.* **3**, 307; Castillo, R., Domínguez, H.C. and Costas, M. (1990) *J. Phys. Chem.* **94**, 8731.
4. Chatwin, P.C. and Allen, C.M. (1985) *Ann. Rev. Fluid Mech.* **17**, 119.
5. Brenner, H. and Edwards, D.A. (1993) *Macrotransport Processes*, Butterworth-Heinemann, Boston.
6. Frankel, I., Mancini, F. and Brenner, H. (1991) *J. Chem. Phys.* **95**, 8636.
7. Brenner, H. (1979) *J. Colloid Interface Sci.* **71**, 189; (1981) **80**, 548.
8. Brenner, H. (1980) *PhysicoChem. Hydrodyn.* **1**, 91; (1982) **3**, 139.
9. Frankel, I. and Brenner, H. (1989) *J. Fluid Mech.* **204**, 97.
10. Dill, L.H. and Brenner, H. (1982) *J. Colloid Interface Sci.* **85**, 101.
11. Chatwin, P.C. (1970) *J. Fluid Mech.* **43**, 321.
12. Gill, W.N. and Sankarasubramanian, R. (1970) *Proc. R. Soc. London*, Ser. A **316**, 341; (1971) A **322**, 101.
13. Maron, V. (1977) *Int. J. Multiphase Flow* **4**, 339.
14. Smith, R. (1981) *J. Fluid Mech.* **105**, 469; (1982) **182**, 447.
15. Young, W.R. and Jones, S. (1991) *Phys. Fluids A* **3**, 1087; Smith, R. (1987) *J. Fluid Mech.* **175**, 201.
16. Pagitsas, M., Nadim, A. and Brenner, H. (1986) *Physica* **135 A**, 533; (1986) *J. Chem. Phys.* **84**, 2801; Pagitsas, M. (1989) *PhysicoChem. Hydrodyn.* **4**, 467.
17. Van Den Broeck, C. (1982) *Physica A* **112**, 343; (1990) **168**, 677.
18. Jou, D., Casas-Vázquez, J. and Lebon, G. (1993) *Extended Irreversible Thermodynamics*, Springer, Berlin; (1988) *Rep. Prog. Phys.* **51**, 1105.
19. Camacho, J. (1993) *Phys. Rev. E* **47**, 1049.
20. Camacho, J. (1993) *Phys. Rev. E* **48**, 1844.
21. de Groot, S.R. and Mazur, P. (1984) *Nonequilibrium Thermodynamics*, Dover, New York.
22. Kintchin, A. (1963) *Mathematical Foundations of Information Theory*, Dover, New York.
23. Camacho, J. (1993) *Phys. Rev. E* **48**, 310.

24. Jou, D., Camacho, J. and Grmela, M. (1991) *Macromolecules* **24**, 3597.
25. Thacker, W.C. (1976) *J. Phys. Oceanog.* **6**, 66.
26. Bacri, J.C., Bouchaud, J.P., Georges, A., Guyon, E., Hulin, J.P., Rakotomalala, N. and Salin, D. (1990), in J.P. Hulin et al. (eds.), *Hydrodynamics of Dispersed Media*, North-Holland, Amsterdam, pp.249.
27. Rigord, P. (1990) Ph.D. thesis, Université de Paris 6.

HIGHER-ORDER TRANSPORT

L.C. WOODS
Mathematical Institute
University of Oxford, England.

1. Introduction

The expansion of the constitutive equations for the heat flux vector $\mathbf{q}$ and viscous stress tensor $\boldsymbol{\pi}$ in power series of the Knudsen number parameter ε, is a standard procedure in the kinetic theory of gases. (The Knudsen number is the ratio of a microscopic scale of length or time to a corresponding macroscopic scale and for local thermodynamic equilibrium to hold, it is necessary that $\varepsilon \ll 1$.) Classical transport theory deals with 'first-order' constitutive equations, i.e. the first term in the power series,

$$\mathbf{q} = \mathbf{q}_1 + \mathbf{q}_2 + \cdots, \qquad \boldsymbol{\pi} = \boldsymbol{\pi}_1 + \boldsymbol{\pi}_2 + \cdots, \tag{1}$$

where $\mathbf{q}_r = O(\varepsilon^r)$ and $\boldsymbol{\pi}_r = O(\varepsilon^r)$. In a simple fluid consisting of monatomic molecules, these equations are

$$\left.\begin{array}{rll} \mathbf{q}_1 & = & -\kappa\nabla T \qquad\qquad\qquad (\kappa \geq 0), \\[2mm] \boldsymbol{\pi}_1 & = & -2\mu\overset{\circ}{\mathbf{e}} \qquad (\overset{\circ}{\mathbf{e}} \equiv \overset{\circ}{\nabla}\mathbf{v}) \qquad (\mu \geq 0), \end{array}\right\} \tag{2}$$

where T is the temperature, $\mathbf{v}$ is the fluid velocity, $\overset{\circ}{\mathbf{e}}$ is the deviator of the rate of strain tensor $\mathbf{e}$, κ is the thermal conductivity and μ is the coefficient of viscosity.

In this article we shall derive expressions for the heat flux vector and viscosity tensor in a one-component gas that are accurate to $O(\varepsilon^2)$. The usual treatment of this problem starts from a kinetic equation for the evolution of the velocity distribution function. Unfortunately the standard kinetic equation, due to Boltzmann, produces spurious terms in $\mathbf{q}_2$ and $\boldsymbol{\pi}_2$. Here we introduce a new 'fluid' approach that avoids advanced kinetic theory.

We then discuss whether or not it is possible to deduce relations for the heat flux vector and the viscosity tensor that are independent of the Knudsen number. This amounts to finding constitutive equations valid in flows

75

J. S. Shiner (ed.), Entropy and Entropy Generation, 75–99.
© 1996 *Kluwer Academic Publishers.*

that can be described as being 'well-displaced' from equilibrium. Finally a phenomenological theory that claims success in this, namely 'extended' irreversible thermodynamics, is briefly described.

2. Second-order Entropy

2.1. RELAXATION TIMES

An essential element in transport is that the average microscopic time interval between successive collisions between molecules—termed the 'collision interval'—and denoted below by τ, is finite and positive. It follows from kinetic theory[1] that μ and κ are proportional to τ;

$$\mu = p\tau_1, \qquad \kappa = \tfrac{5}{2}c_v\mu = \tfrac{5}{2}Rp\tau_2 \qquad (\tau_2 = \tfrac{3}{2}\tau_1), \qquad (3)$$

where by τ_i, $i = 1, 2, \ldots$ we mean microscopic time intervals of the same order as τ, typically $\alpha\tau$ where α ranges from unity to three or four. In (3) p is the thermodynamic pressure and $R = k_B/m$ is the gas constant, where k_B is Boltzmann's constant and m is the molecular mass. In the first-order theory, τ_1 is the 'relaxation' time for momentum transport, i.e. the average time it takes a molecule initially moving in a given direction to lose its momentum in that direction due to collisions. And τ_2 is a similar relaxation time for particle energy. Other times, $\tau_3, \tau_4, \ldots$, will be introduced later. Their relation to τ_1 can be determined only with the aid of kinetic theory.

It is a well-known result of kinetic theory that for molecules repelling each other with a force proportional to an inverse distance power law,

$$\mu \propto T^s \qquad (\tfrac{1}{2} \leq s \leq \tfrac{5}{2}), \qquad (4)$$

where the upper limit for s corresponds to the Coulomb force acting between charged particles, and the lower limit to collisions between 'hard-core' molecules.

2.2. SPECIFIC ENTROPY

The specific entropy s can be expanded in the Knudsen number series

$$s = s_0 + s_2 + O(\varepsilon^3) \qquad (s_2 < 0), \qquad (5)$$

where $\qquad\qquad s_0 = c_p \ln T - R \ln p + \text{const.},$

c_p being the specific heat at constant pressure. The absence of s_1 is a general result that follows from the convexity of the entropy function. An

---

[1]For an accurate, mean-free-path treatment of kinetic theory, see [1], Chap. 3.

expression for s_2 can be found from σ_1, the first-order entropy production rate.

Imagine the local thermodynamic system, P say, to be abruptly isolated from its surroundings, then after a relaxation time τ, the gradients driving $\boldsymbol{\pi}_1$ and $\mathbf{q}_1$ will disappear and the entropy will have increased from its initial value of $(s_0 + s_2)$ to s_0. As σ_1 falls from its initial value to zero during the relaxation, the average rate of entropy increase is half the inital value of σ_1. Hence

$$\rho s_2 = -\tfrac{1}{2}\tau\sigma_1 \qquad (\tau > 0). \tag{6}$$

This is not accurate, since σ_1 consists of two components that relax at slightly different rates. It is a standard result that ([2], p. 213)

$$\sigma_1 = -T^{-1}(\boldsymbol{\pi}_1 : \overset{\circ}{\mathbf{e}} + \mathbf{q}_1 \cdot \nabla \ln T), \tag{7}$$

the first term of which relaxes in a time τ_1 and the second in a time τ_2. Therefore (6) is replaced by

$$\rho s_2 = \frac{1}{2T}(\tau_1\,\boldsymbol{\pi}_1 : \overset{\circ}{\mathbf{e}} + \tau_2\,\mathbf{q}_1 \cdot \nabla \ln T). \tag{8}$$

By (1) and (2) the expression for s_2 can be written in either of the forms:

$$s_2 \;=\; -R\tau_1^2(\overset{\circ}{\mathbf{e}} : \overset{\circ}{\mathbf{e}} + \tfrac{45}{16}R\nabla T \cdot \nabla \ln T), \tag{9}$$

$$s_2 \;=\; -(4\rho p T)^{-1}(\boldsymbol{\pi}_1 : \boldsymbol{\pi}_1 + \tfrac{4}{5}\mathbf{q}_1 \cdot \mathbf{q}_1/RT). \tag{10}$$

A cardinal principle of thermodynamics is that entropy is a function of state, so it is evident from these equations that we should extend state space, to include either the temperature and fluid velocity gradients, or the fluxes $\boldsymbol{\pi}_1$ and $\mathbf{q}_1$. As will become evident later, the second option is preferable; equation (10) proves to be a better representation of the physical mechanism of transport than (9).

2.3. GENERALIZED GIBBS' EQUATION

The usual form for Gibbs' equation is

$$T\,ds_0 = du + p\,d\rho^{-1},$$

where u is the internal specific energy and ρ is the density. To second-order accuracy this is generalized to

$$T\,ds = du + p\,d\rho^{-1} + T\,ds_2,$$

where from (10) and $u = c_v T$, we can express $T\, ds_2$ in terms of the differentials dT, $d\rho^{-1}$, $d\boldsymbol{\pi}_1$ and $d\mathbf{q}_1$. Thus

$$T\, ds = c_v^*\, dT + p^*\, d\rho^{-1} - \frac{\boldsymbol{\pi}_1 : d\boldsymbol{\pi}_1}{2R\rho^2 T} - \frac{2\mathbf{q}_1 \cdot d\mathbf{q}_1}{5R^2\rho^2 T^2}, \qquad (11)$$

where

$$c_v^* \equiv c_v + \frac{\boldsymbol{\pi}_1 : \boldsymbol{\pi}_1}{2R\rho^2 T^2} + \frac{3\mathbf{q}_1 \cdot \mathbf{q}_1}{5R^2\rho^2 T^3},$$

and

$$p^* \equiv p - \frac{\boldsymbol{\pi}_1 : \boldsymbol{\pi}_1}{2R\rho T} - \frac{2\mathbf{q}_1 \cdot \mathbf{q}_1}{5R^2\rho T^2}.$$

Notice that (11) could be rearranged in the form $ds = \theta^{-1}\, du + \cdots$, where $\theta \equiv (c_v/c_v^*)T$ *might* be regarded as being a generalization of the temperature [3]. But it is evident that this modification is not unique. Although it is true that the choice of $\boldsymbol{\pi}_1$ and $\mathbf{q}_1$ as independent variables has some advantages, it is not mandatory, and had we adopted instead $\mathbf{G} \equiv \tau_1 R^{1/2}\overset{\circ}{\mathbf{e}}$ and $\boldsymbol{\xi} \equiv \tau_1 R \nabla T^{1/2}$, the coefficients of dT and $d\rho^{-1}$ would have remained as c_v and p. However, new definitions of a variable that do not entail its original properties carry the risk of error.

By the standard stability condition applied to a local thermodynamic system P of volume $d\tau$, constant mass $\rho\, d\tau$ and entropy $s\rho\, d\tau$, we obtain

$$d^2 s \leq 0 \qquad \text{(for stability)}. \qquad (12)$$

We shall use this constraint to determine limiting values for the fluxes $\mathbf{q}_1$ and $\boldsymbol{\pi}_1$, and hence upper limits for the magnitude of the Knudsen number.

2.4. STABILITY LIMIT ON KNUDSEN NUMBERS

To investigate the stability of the thermodynamic system defined by the state variables T, $v\ (\equiv \rho^{-1})$ and $q\ (\equiv |\mathbf{q}_1|)$, we write (11) in the form

$$ds = A(T, v, q)\, dT + B(T, v, q)\, dv + C(T, v, q)\, dq, \qquad (13)$$

where (for a monatomic gas)

$$A = \frac{3R}{2T} + \frac{3q^2}{5R^2\rho^2 T^4}, \quad B = R\rho - \frac{2q^2}{5R^2\rho T^3}, \quad C = -\frac{2q}{5R^2\rho^2 T^3}. \qquad (14)$$

Adopting matrix notation and using subscripts to denote partial derivatives, we find from (13) that

$$d^2 s = [dT\ dv\ dq]\, \mathbf{D}\, \{dT\ dv\ dq\}, \qquad (15)$$

where

$$D = \begin{bmatrix} A_T & A_v & A_q \\ B_T & B_v & B_q \\ C_T & C_v & C_q \end{bmatrix}, \tag{16}$$

and $\{\cdots\}$ denotes a column matrix.

For stability, it follows from (12) that the quadratic form in (15) must be negative definite. This is the case if A_T, B_v, C_q and the determinant $|D|$ are negative and the principal minor $(A_T B_v - A_v B_T)$ is positive. It follows from (14) that

$$D = \begin{bmatrix} -\dfrac{3}{2T^2}(R + 4Tq^2h) & 3\rho q^2 h & 3qh \\ 3\rho q^2 h & -\rho^2(R + Tq^2h) & -2\rho Tqh \\ 3qh & -2\rho Tqh & -Th \end{bmatrix},$$

where $h \equiv 2/(5R^2\rho^2T^4)$. Each term in the leading diagonal is negative, as required for stability. The principal minor obtained by omitting the last row and column, is positive if

$$(R + 4x)(R + x) - 6x^2 > 0 \qquad (x \equiv Tq^2h).$$

This requires that $q < q_*$, where $q_* \approx 2.59\rho(RT)^{3/2}$. Now

$$|q_1| = |\tfrac{5}{2}Rp\tau_2\nabla T| \sim 1.25pC\lambda/\mathcal{L} \qquad (C = (2RT)^{1/2}, \quad \lambda = \tau_2 C),$$

where λ is the mean free path and $\mathcal{L}$ is the macroscopic length scale. It follows that the critical value of the Knudsen number, $\varepsilon = \lambda/\mathcal{L}$, is $\varepsilon_* \approx 1.46$, but, as we shall see shortly, a much smaller value is required to ensure that $|D|$ is negative.

Evaluating the determinant we get

$$|D| = -\dfrac{3h\rho^2}{2T}(R^2 + 7xR - 24x^2),$$

which is negative provided x does not exceed $0.397R$. This yields the constraint $q < q_* \approx 0.996\rho(RT)^{3/2}$, which corresponds to a critical Knudsen number of $\varepsilon_* = 0.56$.

A similar treatment, in which the viscous stress tensor is retained and the heat flux omitted, gives a critical value for $\|\pi_1\|$ of $0.722p$. Therefore we have $2p\|\tau_1 \overset{\circ}{e}\| = 2p\varepsilon_* = 0.722p$, giving $\varepsilon_* = 0.36$.

Summing up these results, we see that the linear theory of transport is inconsistent with stable flows for values of the Knudsen number above

about a third:
$$\varepsilon < \varepsilon_* \approx \tfrac{1}{3}. \tag{17}$$

Of course the accuracy of the linear theory will decrease the closer ε approaches this limit. How much the value of ε_* can be raised by including terms second-order in ε in the constitutive equations is difficult to judge. Since the linear term plays a dominant rôle in transport, it is unlikely that higher-order terms would increase ε_* very much; a limiting value greater than unity would seem improbable.

3. Second-order Heat Flux

3.1. PHYSICAL MECHANISMS

We start by reminding the reader of the physical nature of the diffusion vector $\mathbf{J}_\phi$ for a property $\phi(\mathbf{r}, t)$. The conservation equation for ϕ reads

$$\frac{\partial}{\partial t}(\rho\phi) + \boldsymbol{\nabla} \cdot (\rho\phi\mathbf{v} + \mathbf{J}_\phi) = S_\phi,$$

where the right-hand side represents a source term. The total flux of ϕ across an element of surface $\mathbf{n}\,dS$ is $\rho\phi\mathbf{v} \cdot \mathbf{n}\,dS + \mathbf{J}_\phi \cdot \mathbf{n}\,dS$. The first term in this expression is the 'convection' of ϕ across the surface element, while the second is the 'diffusive' flux of ϕ across the element.

Diffusion is the transport of a property by the purely *random* component of molecular motion. It is essential to distinguish between diffusion and convection. We can do this by noting that the convection term is dependent on the choice of reference frame in which the velocity is measured, whereas the diffusion term is not. 'Frame-indifference' is the essential property that distinguishes diffusion from convection. And it should be remembered that not only must the local thermodynamic system P be moving with the fluid velocity $\mathbf{v}$, to eliminate all ordered particle motions from P, it must also have the *acceleration*, $D\mathbf{v}$, and *spin*, $\boldsymbol{\Omega} \equiv \tfrac{1}{2}\boldsymbol{\nabla}\times\mathbf{v}$, of the fluid element in question. Note that the rate of change of $\mathbf{q}$ in P is given by the Jaumann time derivative

$$D^*\mathbf{q} = D\mathbf{q} - \boldsymbol{\Omega}\times\mathbf{q} = \frac{\partial\mathbf{q}}{\partial t} + \mathbf{v}\cdot\boldsymbol{\nabla}\mathbf{q} - \boldsymbol{\Omega}\times\mathbf{q}. \tag{18}$$

It follows that $\mathbf{J}_\phi$ is convected with the fluid and in particular that this is true of $\mathbf{q}$ and $\boldsymbol{\pi}$.

It is important to distinguish between *energy* flux, which does not necessarily involve collisions and *heat* flux, which does. With the former, P's ambient conditions play no part, whereas collisions turn energy flux into heat. Both fluxes are unfortunately represented by the single vector $\mathbf{q}$, although the context should make clear which phenomenon is being considered.

The fluid pressure gradient ∇p contributes to the acceleration of P, and therefore, viewed from P, has no effect on the diffusion of molecules from P. For this reason ∇p can not appear in the formula for the heat flux vector $\mathbf{q}$, which constraint is independent of the order in ε of the term in question. A more compelling argument against the appearance of second-order terms like $\nabla \mathbf{v} \cdot \nabla p$ and $\nabla \cdot \boldsymbol{\pi}$ in $\mathbf{q}$ is that they would yield a heat flux in *isothermal* conditions, i.e. in conditions when the internal energy is everywhere the same. Any transfer of energy would immediately result in heat flowing *up* a small temperature gradient, which by Clausius' form of the second law is not possible in a simple fluid.

In view of the relation $p = R\rho T$, we must also rule out $\nabla \rho$ from the list of gradients able to generate a heat flux in a simple fluid. We are left with ∇T. The only second-order vectorial terms that we can form using ∇T and the operators D^* and ∇, are $D^* \nabla T$ and $\nabla \mathbf{v} \cdot \nabla T$. Thus we know what terms can appear in $\mathbf{q}_2$, and it remains to determine the mechanism that produces them.

The first-order flux $\mathbf{q}_1$ has a special rôle in the theory. Being proportional to ∇T, it depends on distant boundary conditions where heat is supplied or removed from the system. With these conditions held constant, $\mathbf{q}_1$ is the 'equilibrium' value of $\mathbf{q} = \mathbf{q}_1 + \mathbf{q}_2 + \cdots$, with $\mathbf{q}_2$ being a perturbation due to time-delays, as we shall next explain.

3.2. PERSISTENCE OF THE RELATIVE HEAT FLUX

When collisions can be said to dominate, i.e. the Knudsen number is quite small, heat passes through the local thermodynamic system P via molecular vibrations that are independent of P's motion. (Of course we are excluding the convection of energy in this reckoning.) In this case $\mathbf{q}$ has its equilibrium value $\mathbf{q}_1$. But if collisions are less frequent, so that the collision interval τ, while remaining smaller than the macroscopic time scale, is significant, the transport of particle energy is influenced by the shearing, spinning and dilatory motions of P. This occurs because the random component of molecular energy is transmitted by particles momentarily in free transit, and the paths of these particles are randomly distributed only when viewed from the system P. (In the near absence of collisions, the vector $\mathbf{q}$ represents *energy* flux rather than heat, and is almost completely embedded in P.) It follows that we need a method of specifying P relative to the flux of heat through it.

Let $\boldsymbol{\lambda}$ denote an infinitesimal vector, about a mean free path in length, that is embedded in the thermodynamic system P. As P accelerates and spins relative to the laboratory frame, so does $\boldsymbol{\lambda}$. This vector therefore 'tags' the system P in which the heat flux $\mathbf{q}$ is measured. Its value is arbitrary.

Thus λ is a generic tag representing the environment P in which the heat flux $\mathbf{q}$ is generated. Since by $\mathbf{q}$ we mean the collisional deposition of energy, there is a delay time of a collision interval τ between changes in P and the response in $\mathbf{q}$. Thus, for a given $\mathbf{q}(t)$, it is necessary to characterize P by the value of λ at a time τ earlier than the present time, t.

Let λ_τ denote the value of λ at time $(t - \tau)$, then we define the *relative heat flux* by

$$\xi \equiv \mathbf{q} \cdot \lambda_\tau \approx \mathbf{q} \cdot (\lambda - \tau D^* \lambda), \qquad (19)$$

the approximate form following from a Taylor expansion. The changes in $\mathbf{q}$ of interest below are those relative to P and they can be deduced from an expression giving the rate of change of ξ. As remarked above, when collisions are infrequent, $\mathbf{q}(t)$ is embedded in $P(t - \tau)$ and since the energy flux is constant between collisions, the scalar ξ is constant; collisions tend to drive ξ towards its equilibrium value of $\xi_1 \equiv \mathbf{q}_1 \cdot \lambda_\tau$.

To simplify the model, we shall lump the collisions into groups, spaced at intervals τ along the trajectory of P, with ξ changing abruptly into ξ_1 at the end of each interval. Of course such synchronous behaviour is unreal, but does not violate the essential physics. A 'continuous' version of the process will be described shortly. We may picture the property ξ as being swept along with the local thermodynamic system P, persisting in value for a time proportional to τ and then being changed to ξ_1 by collisions.

Suppose that at a time t, collisions have momentarily restored ξ to its equilibrium value $\xi_1(t)$, then in the absence of further collisions, its magnitude will persist until time $(t + \tau)$, when it becomes the relative flux $\xi(t + \tau)$ At this stage collisions change it to $\xi_1(t + \tau)$. Thus

$$\xi_1(t) = \xi(t + \tau) = \xi(t) + \tau D\xi + O(\varepsilon^3), \qquad (20)$$

where D is the usual material derivative. (For a scalar operand $D^* = D$.) As ξ is proportional to $\mathbf{q}$, it is an $O(\varepsilon)$ term; the operator τD is $O(\varepsilon)$, making $\tau D\xi$ a second-order term.

There is the possibility that the $O(\varepsilon^3)$ term is zero, in which case (20) reduces to the simple relaxation law

$$D\xi = \frac{1}{\tau}(\xi_1 - \xi) \qquad (D \equiv \frac{\partial}{\partial t} + \mathbf{v} \cdot \boldsymbol{\nabla}). \qquad (21)$$

This equation has the merit of representing a continuous mechanism for the collisional changes in ξ, to which topic we shall return in §5.1.

The situation is analogous to that of the velocity distribution function $f(\mathbf{r}, \mathbf{c}, t)$ in the kinetic theory of gases (see [1], pp. 92–4). In that theory, to a good approximation, the number of particles $f\,d\nu$ in an element of 6-D phase space $d\nu$, is equal to the number in the equilibrium distribution $f_0\,d\nu$ a collision interval earlier. Of course the number is continuously altered by the collisions, but if they are synchronous, numbers are conserved over a collison interval. Since the volume element $d\nu$—which corresponds to the tag λ—does not change along the trajectory, we get $f_0(\mathbf{r}, \mathbf{c}, t) \approx f(\mathbf{r} + \tau\mathbf{c}, \mathbf{c} + \tau\mathbf{a}, t+\tau)$, where $\mathbf{c}$ and $\mathbf{a}$ are the particle velocity and acceleration measured in P. This relation leads to the well-known 'BGK' kinetic equation,

$$\mathbb{D}f = \frac{1}{\tau}(f_0 - f), \qquad (22)$$

where $\mathbb{D}$ is the convective rate of change in 6-D phase space.

The error in the second form of ξ given in (19) is $O(\varepsilon^3)$, and therefore (20) can be written

$$\tau D^*\{\mathbf{q} \cdot (\lambda - \tau D^*\lambda)\} + \mathbf{q} \cdot (\lambda - \tau D^*\lambda) = \mathbf{q}_1 \cdot (\lambda - \tau D^*\lambda) + O(\varepsilon^3),$$

or $\qquad (\tau D^*\mathbf{q} + \mathbf{q}) \cdot \lambda = \mathbf{q}_1 \cdot (\lambda - \tau D^*\lambda) + O(\varepsilon^3).$ $\qquad (23)$

In this equation τ will have slightly different values for each of the properties undergoing relaxation, a point that will become clearer below. In the rest of this article, we shall use τ to indicate a variety of relaxation times; when we wish to be specific about τ, we shall add subscripts, as in (3).

3.3. FINAL EXPRESSION FOR THE SECOND-ORDER HEAT FLUX

Consider a short material line element $d\mathbf{x}$, namely one that is convected with the fluid, then the rate of change of $d\mathbf{x}$ is just the difference between the velocities at the two end of the element. Hence $D(d\mathbf{x}) = d\mathbf{x} \cdot \nabla\mathbf{v}$. In the mechanism described above, λ is a vector like $d\mathbf{x}$. Therefore, since

$$\nabla\mathbf{v} = \mathbf{e} - \boldsymbol{\Omega}\times\mathbf{1} = \overset{\circ}{\mathbf{e}} - \boldsymbol{\Omega}\times\mathbf{1} + \tfrac{1}{3}\nabla\cdot\mathbf{v}\,\mathbf{1}, \qquad (24)$$

we get $D\lambda = \lambda \cdot \nabla\mathbf{v} = \lambda \cdot \mathbf{e} - \lambda\times\boldsymbol{\Omega}$. Thus

$$D\lambda - \boldsymbol{\Omega}\times\lambda \equiv D^*\lambda = \lambda \cdot \mathbf{e} \qquad (25)$$

is the rate of change of λ in the frame of the thermodynamic system P. Therefore, as $\mathbf{e}$ is a symmetric tensor,

$$\lambda - \tau D^*\lambda = \lambda - \tau\lambda \cdot \mathbf{e} = (\mathbf{1} - \tau\mathbf{e}) \cdot \lambda.$$

Hence (23) may be written

$$(\tau D^*\mathbf{q} + \mathbf{q}) \cdot \lambda = \mathbf{q}_1 \cdot (\mathbf{1} - \tau\mathbf{e}) \cdot \lambda + O(\varepsilon^3),$$

and as this holds for all choices of λ, we arrive at

$$\tau D^* \mathbf{q} + \mathbf{q} = \mathbf{q}_1 - \tau \mathbf{e} \cdot \mathbf{q}_1 + O(\varepsilon^3). \tag{26}$$

Now $\mathbf{q} = \mathbf{q}_1 + \mathbf{q}_2 + O(\varepsilon^3)$, so (26) yields

$$\mathbf{q}_2 = -\tau D^* \mathbf{q}_1 - \tau \mathbf{e} \cdot \mathbf{q}_1 = -\tau D^* \mathbf{q}_1 - \tau \overset{\circ}{\mathbf{e}} \cdot \mathbf{q}_1 - \tfrac{1}{3}\tau \mathbf{q}_1 \nabla \cdot \mathbf{v}.$$

This expression contains all the second-order terms that we predicted earlier from physical considerations.

At this stage we distinguish between the terms in $\mathbf{q}_2$ by allowing them to relax at slightly different rates. Hence

$$\mathbf{q}_2 = -\{\tau_3 D^* \mathbf{q}_1 + \tau_4 \overset{\circ}{\mathbf{e}} \cdot \mathbf{q}_1 + \tau_5 \mathbf{q}_1 \nabla \cdot \mathbf{v}\}. \tag{27}$$

The relationship between the times τ_3, τ_4, τ_5 and τ_1 can be found only from kinetic theory (see §6.1).

The second right-hand term in (27) *could* be expressed as $\tau_4 \boldsymbol{\pi}_1 \cdot \mathbf{q}_1/2\mu$, but there is no physical justification for this. Here the deviator of $\mathbf{e}$ occurs because of the environmental changes in λ and not in the rôle of a state variable (when it would be better changed into $\boldsymbol{\pi}_1$). One of the confusing aspects of second-order transport is that $\mathbf{e}$ has two functions, which must be distinguished.

4. Second-Order Viscous Stress Tensor

4.1. CONVECTION OF VISCOUS FORCE

A similar approach to that given above for $\mathbf{q}_2$ can be used to find an expression for the second-order viscous stress tensor $\boldsymbol{\pi}_2$. The persisting physical property is $\xi = \lambda_{2\tau} \cdot \boldsymbol{\pi} \cdot \lambda_{1\tau}$, where now two tags are required to characterize P. By (25)

$$\xi = (\lambda_2 - \tau \lambda_2 \cdot \mathbf{e}) \cdot \boldsymbol{\pi} \cdot (\lambda_1 - \tau \lambda_1 \cdot \mathbf{e}) + O(\varepsilon^3),$$

or
$$\xi = \lambda_2 \cdot \boldsymbol{\pi} \cdot \lambda_1 - \tau \lambda_2 \cdot (\mathbf{e} \cdot \boldsymbol{\pi} + \boldsymbol{\pi} \cdot \mathbf{e}) \cdot \lambda_1 + O(\varepsilon^3).$$

The argument leading to (20) also applies with the present expression for ξ. Hence

$$\tau D(\lambda_2 \cdot \boldsymbol{\pi} \cdot \lambda_1) + \lambda_2 \cdot \boldsymbol{\pi} \cdot \lambda_1 - \tau \lambda_2 \cdot (\mathbf{e} \cdot \boldsymbol{\pi} + \boldsymbol{\pi} \cdot \mathbf{e}) \cdot \lambda_1$$

$$= \lambda_2 \cdot \boldsymbol{\pi}_1 \cdot \lambda_1 - \tau \lambda_2 \cdot (\mathbf{e} \cdot \boldsymbol{\pi}_1 + \boldsymbol{\pi}_1 \cdot \mathbf{e}) \cdot \lambda_1 + O(\varepsilon^2),$$

where for reasons to be given below, we have reduced the order of the error term.

As ξ is a scalar, we can replace D by D^* and by (25) reduce the equation to

$$\lambda_2 \cdot [\tau D^* \pi + \pi] \cdot \lambda_1 = \lambda_2 \cdot [\pi_1 - \tau (\mathbf{e} \cdot \pi_1 + \pi_1 \cdot \mathbf{e})] \cdot \lambda_1 + O(\varepsilon^2).$$

On using the relation

$$\mathbf{e} \cdot \pi_1 + \pi_1 \cdot \mathbf{e} = 2 \overset{\circ}{\mathbf{e}} \cdot \pi_1 + \tfrac{2}{3} \mathbf{e} : \pi_1 \mathbf{1} + \tfrac{2}{3} \pi_1 \nabla \cdot \mathbf{v},$$

and the fact that the tags are arbitrary, we arrive at

$$\tau D^* \pi + \pi = \pi_1 - 2\tau \overset{\circ}{\mathbf{e}} \cdot \pi_1 - \tfrac{2}{3} \tau \pi_1 \nabla \cdot \mathbf{v} + O(\varepsilon^2), \qquad (28)$$

where, because π_1 is a deviator, the term containing the unit tensor is omitted. In these equations the symbol $\circ$ denotes a deviator, and the Jaumann derivative of a tensor like π has the form

$$D^* \pi = D\pi - \Omega \times \pi + \pi \times \Omega.$$

The temptation to express $-2\tau \overset{\circ}{\mathbf{e}} \cdot \pi_1$ as $\pi_1 \cdot \pi_1 / p$ must be resisted, since the latter form has no relation to the physical origin of the term.

4.2. REMOVING THE FLUID ACCELERATION

Whereas (27) contained all the expected terms, with the viscous stress tensor the physical situation is not so clear. Possible additional $O(\varepsilon^2)$ terms contain $\nabla T \nabla T$ and $\nabla^2 T$, but at least we do not need to consider $\nabla p \nabla T$, which represents a coupling between a fluid *force* and a temperature gradient. The argument is the same as that used in the fourth paragraph of §3.1—viewed from the accelerating thermodynamic element P, the force $-\rho^{-1} \nabla p$ cannot be 'felt' by the molecules, therefore cannot contribute to any form of diffusion.

The equation of fluid motion can be written

$$D\mathbf{v} = -\rho^{-1} \nabla p + \mathbf{F} + O(\varepsilon).$$

Taking its gradient, we have

$$\nabla D\mathbf{v} = D(\nabla \mathbf{v}) + \nabla \mathbf{v} \cdot \nabla \mathbf{v} = \frac{1}{\rho^2} \nabla \rho \nabla p - \frac{1}{\rho} \nabla \nabla p + \nabla \mathbf{F} + O(\varepsilon).$$

From (24) we find that in the spinning system P ($\Omega = 0$), correct to first order in ε,

$$\Phi \equiv D^* \overset{\circ}{\mathbf{e}} + \overset{\overline{\circ\quad\circ}}{\mathbf{e} \cdot \mathbf{e}} + \tfrac{2}{3} \overset{\circ}{\mathbf{e}} \nabla \cdot \mathbf{v} = \frac{1}{\rho^2} \overline{\nabla \rho \nabla p} - \frac{1}{\rho} \overline{\nabla \nabla p} + \overset{\circ}{\nabla \mathbf{F}}. \qquad (29)$$

Now $\boldsymbol{\Phi}$ is the gradient of the fluid acceleration as seen in the thermodynamic system P. It applies to the collective motion of the particles, that is each molecule is accelerated by the *same* force and therefore $\boldsymbol{\Phi}$ cannot contribute to diffusion. It needs to be removed from the right-hand side of (28).

First consider the physical meaning of $\boldsymbol{\lambda}_2 \cdot \boldsymbol{\pi} \cdot \boldsymbol{\lambda}_1$, where $\boldsymbol{\lambda}_1$ and $\boldsymbol{\lambda}_2$ are a mean free path in length. We shall introduce unit vectors $\hat{\boldsymbol{\lambda}}_1$ and $\hat{\boldsymbol{\lambda}}_2$ and write $\boldsymbol{\lambda}_1 = \tau \mathcal{C} \hat{\boldsymbol{\lambda}}_1$, $\boldsymbol{\lambda}_2 = \tau \mathcal{C} \hat{\boldsymbol{\lambda}}_2$, where $\mathcal{C}$ is the most probable molecular speed, $(2RT)^{\frac{1}{2}}$. Now $\boldsymbol{\lambda}_2 \cdot \boldsymbol{\pi} \cdot \boldsymbol{\lambda}_1 = \hat{\boldsymbol{\lambda}}_2 \cdot \boldsymbol{\pi} \cdot d\mathbf{S}$, where $d\mathbf{S} \equiv \tau^2 \mathcal{C}^2 \hat{\boldsymbol{\lambda}}_1$ is the element of area having $\hat{\boldsymbol{\lambda}}_1$ as its unit normal. This allows us to interpret $\boldsymbol{\lambda}_2 \cdot \boldsymbol{\pi} \cdot \boldsymbol{\lambda}_1$ as the component along $\hat{\boldsymbol{\lambda}}_1$ of the surface force acting on one face $d\mathbf{S}$ of a cubical thermodynamic system P of mass $\rho \tau^3 \mathcal{C}^3$.

From its definition $\boldsymbol{\Phi} \cdot \boldsymbol{\lambda}_1$ is the change in the acceleration over the length of the vector $\boldsymbol{\lambda}_1$, which passes through P from its face $d\mathbf{S}$ to the opposite side. Hence $\rho \tau^3 \mathcal{C}^3 \boldsymbol{\Phi} \cdot \boldsymbol{\lambda}_1$ is the force acting on P along $\hat{\boldsymbol{\lambda}}_1$, and $\hat{\boldsymbol{\lambda}}_2 \cdot \rho \tau^3 \mathcal{C}^3 \boldsymbol{\Phi} \cdot \boldsymbol{\lambda}_1 = \boldsymbol{\lambda}_2 \cdot (\rho \tau^2 \mathcal{C}^2 \boldsymbol{\Phi}) \cdot \boldsymbol{\lambda}_1$ is its component in the $\hat{\boldsymbol{\lambda}}_2$-direction. This force cannot contribute to the diffusion of particle momentum. We have therefore identified one of the remaining $O(\varepsilon^2)$ terms in (28) to be equal to $-\rho \tau^2 \mathcal{C}^2 \boldsymbol{\Phi} = -2p\tau^2 \boldsymbol{\Phi} = -2\mu\tau \boldsymbol{\Phi}$.

4.3. VISCOUS STRESS DUE TO HEAT FLUX GRADIENT

There remain the possible terms $\nabla T \nabla T$ and $\nabla^2 T$. From (3) and (4) we get $\nabla \kappa = s\kappa \nabla T / T$, from which it follows that $-\tau_7 \nabla \mathbf{q}$ ($\tau_7 \geq 0$) contains both these terms—albeit in a particular combination. The physical basis for such a term is as follows.

We take the special case of flow in which there is no velocity shear, there is a temperature gradient along the axis $-OY$, and the resulting first-order heat flux along OY is sheared in a perpendicular direction, along OX. The existence of a $\nabla \mathbf{q}$ term in $\boldsymbol{\pi}_2$ can be deduced by the following reasoning. The gas may be divided into two superimposed pseudo-fluids, F_f containing the faster than average molecules and F_s containing the slower molecules. On average the fast molecules move along OY from the hot to the cold region, while the slow molecules move, again on average, along $-OY$. Transverse molecular motion along OX will transmit a shear force parallel to $-OY$ in F_f and parallel to OY in F_s. Since the fast molecules have longer mean free paths than the slow ones, it follows that the shear force in F_f is larger than that in F_s, i.e. there is net force transmitted along $-OY$. By analogy with the case of fluid velocity shear, this force will be proportional to dq/dx, and to the average collision interval, τ_2. Thus this heuristic argument yields a force per unit area of $\hat{\mathbf{x}} \cdot \boldsymbol{\pi}_2 \propto -\tau_2 (dq/dx) \hat{\mathbf{y}}$, as required.

4.4. EXPRESSION FOR THE VISCOUS STRESS TENSOR

With the additional $O(\varepsilon^2)$ terms derived in §§4.2 and 4.3, equation (28) now reads:

$$\tau D^* \boldsymbol{\pi} + \boldsymbol{\pi} = \boldsymbol{\pi}_1 - \tau\{2\overline{\overset{\circ}{\mathbf{e}} \cdot \boldsymbol{\pi}_1} + \tfrac{2}{3}\boldsymbol{\pi}_1 \boldsymbol{\nabla} \cdot \mathbf{v}\} - \tau \overline{\boldsymbol{\nabla} \overset{\circ}{\mathbf{q}}_1}$$

$$-2\mu\tau\{D^*\overset{\circ}{\mathbf{e}} + \overline{\overset{\circ}{\mathbf{e}} \cdot \overset{\circ}{\mathbf{e}}} + \tfrac{2}{3}\overset{\circ}{\mathbf{e}}\boldsymbol{\nabla} \cdot \mathbf{v}\} + O(\varepsilon^3),$$

or $\qquad \tau D^* \boldsymbol{\pi} + \boldsymbol{\pi} = \boldsymbol{\pi}_1 + 2\mu\tau\left(\overline{\overset{\circ}{\mathbf{e}} \cdot \overset{\circ}{\mathbf{e}}} - D^*\overset{\circ}{\mathbf{e}}\right) - \tau \overline{\boldsymbol{\nabla} \overset{\circ}{\mathbf{q}}} + O(\varepsilon^3),$ $\qquad$ (30)

where the error term can now be reduced to third order.

To obtain an expression for $\boldsymbol{\pi}_2$, we substitute $\boldsymbol{\pi} = \boldsymbol{\pi}_1 + \boldsymbol{\pi}_2 + O(\varepsilon^3)$ into (30). From (4) and the energy equation,

$$D\mu = s\mu D \ln T = -\tfrac{2}{3}s\mu\boldsymbol{\nabla} \cdot \mathbf{v} + O(\varepsilon), \qquad (31)$$

whence $\qquad D^*\boldsymbol{\pi}_1 = -2D^*(\mu\overset{\circ}{\mathbf{e}}) = -\tfrac{2}{3}s\,\boldsymbol{\pi}_1 \boldsymbol{\nabla} \cdot \mathbf{v} - 2\mu D^*\overset{\circ}{\mathbf{e}}.$

Therefore (30) yields

$$\boldsymbol{\pi}_2 = -\tau_6\,\overline{\overset{\circ}{\mathbf{e}} \cdot \boldsymbol{\pi}_1} - \tau_7\,\overline{\boldsymbol{\nabla} \overset{\circ}{\mathbf{q}}_1} - \tau_8 \boldsymbol{\pi}_1 \boldsymbol{\nabla} \cdot \mathbf{v}, \qquad (32)$$

where the separate terms are now allowed to relax at slightly different rates.

The absence of $D^*\boldsymbol{\pi}_1$ from (32) appears to be in conflict with a famous relation due to Maxwell (see §7.2)—a relation that is one of the linchpins of what is termed 'extended irreversible thermodynamics'.

4.5. COMPARISON WITH KINETIC THEORY

Our second-order transport equations are (27) and (32):

$$\left. \begin{aligned} \mathbf{q}_2 &= -\tau_3 D^*\mathbf{q}_1 - \tau_4 \overset{\circ}{\mathbf{e}} \cdot \mathbf{q}_1 - \tau_5 \mathbf{q}_1 \boldsymbol{\nabla} \cdot \mathbf{v}, \\[2mm] \boldsymbol{\pi}_2 &= -\tau_6\,\overline{\overset{\circ}{\mathbf{e}} \cdot \boldsymbol{\pi}_1} - \tau_7\,\overline{\boldsymbol{\nabla} \overset{\circ}{\mathbf{q}}_1} - \tau_8 \boldsymbol{\pi}_1 \boldsymbol{\nabla} \cdot \mathbf{v}. \end{aligned} \right\} \qquad (33)$$

When these formulae are derived from kinetic theory ([1], pp. 175–7), one finds that the microscopic times are related by

$$\left. \begin{array}{llll} \tau_3 = 2\tau_2, & \tau_4 = (\tfrac{14}{3} + \tfrac{8}{15}s)\tau_2, & \tau_5 = (3 + \tfrac{2}{3}s)\tau_2, \\[2mm] \tau_6 = 3\tau_1, & \tau_7 = \tfrac{4}{5}\tau_1, & \tau_8 = (3 - \tfrac{2}{3}s)\tau_1. \end{array} \right\} \qquad (34)$$

Since $\tau_2 = \frac{3}{2}\tau_1$, they can all be expressed in terms of $\tau_1 = \mu/p$.

It should be pointed out that (33) are *not* the well-known Burnett equations. Those have additional terms involving ∇p, and will be presented in §7.2. Our equations have been shown to give better agreement with experiments with ultrasonic waves and with strong shock waves than the Burnett equations (see [1], pp. 181–9).

5. Generalized Heat Flux Law

5.1. INTEGRAL FORM FOR THE HEAT FLUX

The derivation of (20) depends on the approximation of adopting the first term in a Taylor expansion. While this could be extended to higher orders, the algebra would soon mask the physics. In this section we shall develop a non-synchronous model of the collisional process $\xi \rightarrow \xi_1$, allowing us to avoid the Taylor expansion.

The relative heat flux, $\xi \equiv \mathbf{q} \cdot \boldsymbol{\lambda}_\tau$ has an equilibrium value ξ_1 determined by conditions external to the local thermodynamic system P. Flux is added to (or subtracted from) P by molecular interactions as it moves to the point $\mathbf{r}, t$. Suppose that P is at the point $\mathbf{r}', t'$, upstream of the current point $\mathbf{r}, t$, then in the range $(t', t' + dt')$, flux is added to P with a probability proportional to dt'/τ. We adopt the model represented in (20), namely that the current value of ξ depends on its upstream equilibrium value, $\xi_1(\mathbf{r}', t')$. The probability of this influence surviving collisions between $(\mathbf{r}', t')$ and $(\mathbf{r}, t)$ is $\exp\{-(t - t')/\tau\}$. Thus (20) is replaced by the integral relation

$$\xi(\mathbf{r}, t) = \int_{-\infty}^{t} e^{-(t-t')/\tau} \xi_1(\mathbf{r}', t')\, dt'/\tau, \tag{35}$$

where
$$\mathbf{r}' = \mathbf{r} - \int_{t'}^{t} \mathbf{v}''\, dt''. \tag{36}$$

It follows from $\xi_1 = \mathbf{q}_1 \cdot \boldsymbol{\lambda}_\tau$ that

$$[\mathbf{q} \cdot \boldsymbol{\lambda}_\tau](\mathbf{r}, t) = -\int_{-\infty}^{t} e^{-(t-t')/\tau} [\kappa \nabla T \cdot \boldsymbol{\lambda}_\tau](\mathbf{r}', t')\, dt'/\tau.$$

In the absence of fluid motion $\boldsymbol{\lambda}_\tau$ is constant, so that

$$\mathbf{q}(\mathbf{r}, t) = -\int_{-\infty}^{t} e^{-(t-t')/\tau}\, \kappa \nabla T(\mathbf{r}', t')\, dt'/\tau. \tag{37}$$

5.2. GENERALIZED FOURIER'S LAW

Differentiating (36) and (35), we get $\mathbf{Dr'} = \mathbf{v} - \mathbf{v} = 0$ and

$$D(\xi) = \frac{1}{\tau}\xi_1 - \frac{1}{\tau}\xi - \int_{-\infty}^{t} e^{-(t-t')/\tau} D(\xi_1(\mathbf{r'}, t')) \, dt'/\tau,$$

where

$$D\xi_1 = \mathbf{Dr'} \cdot \frac{\partial \xi_1'}{\partial \mathbf{r'}} = 0.$$

Hence (cf. (21))

$$D\xi = \frac{1}{\tau}(\xi_1 - \xi), \tag{38}$$

i.e. ξ is driven by collisions towards its equilibrium value ξ_1, in a 'relaxation' time τ, and at a rate proportional to the displacement $(\xi_1 - \xi)$. In this form, it is a physically tenable empirical equation. This relaxation process is of the type first introduced by Maxwell [4] for the viscous stress (see §7.2).

It follows from (19), (25) and (38) that, on allowing for variations in τ,

$$\tau_3 D^* \mathbf{q} + \mathbf{q} = -\kappa \nabla T + \tau_4 \kappa \nabla T \cdot \overset{\circ}{\mathbf{e}} + \tau_5 \kappa \nabla T \, \nabla \cdot \mathbf{v} + O(\varepsilon^3). \tag{39}$$

When fluid motions are absent, or can be ignored, (39) reduces to

$$\tau_3 \frac{\partial \mathbf{q}}{\partial t} + \mathbf{q} = -\kappa \nabla T, \tag{40}$$

where the error term, which is associated with the fluid motion (see (19)), has now disappeared. This constitutive relation was introduced by Cattaneo [5] in order to give temperature perturbations a *finite* velocity of propagation, it being supposed that this singular result is a consequence of Fourier's simpler constitutive relation. It is not difficult to show that for $\varepsilon \equiv \omega\tau < 1$, where ω is the wave frequency, thermal 'waves' are evanescent and propagate only a very short distance and with a finite velocity. Only if one accepts the invalid limit $\varepsilon \to \infty$ is there any chance of a different result. One finds that in this case the phase velocity is $v_{p\infty} = (\kappa/(\rho c_v \tau_3))^{\frac{1}{2}}$. In the limit $\tau_3 \to 0$, the phase velocity remains finite because there is a fixed relationship between the collision interval τ_2 in κ and τ_3.

It is important to remember that, although we appear to have avoided the Knudsen number constraint, for the reasons given in §1.4, this condition, namely

$$\varepsilon \equiv \left| \frac{\tau_3}{|\mathbf{q}|} \frac{\partial |\mathbf{q}|}{\partial t} \right| < \varepsilon_* < 1,$$

remains essential to the theory.

6. Entropy Production Rate

6.1. GENERAL FORMULAE

The first law of thermodynamics for a fluid comprised of structureless, monatomic molecules can be arranged in the form

$$\mathrm{D}s_0 + \boldsymbol{\nabla} \cdot \left(\frac{\mathbf{q}}{T}\right) = -\frac{1}{T^2}\boldsymbol{\pi} : \overset{\circ}{\mathbf{e}} - \frac{\mathbf{q}}{T} \cdot \boldsymbol{\nabla}T,$$

where
$$T\mathrm{D}s_0 = \mathrm{D}u + p\mathrm{D}\rho^{-1}.$$

Let $\mathbf{J} = \mathbf{q}/T + \mathbf{J}_2 + \ldots$ denote the total entropy flux vector, then

$$\rho\mathrm{D}s + \boldsymbol{\nabla} \cdot \mathbf{J} = \sigma, \qquad \rho\mathrm{D}s_2 + \boldsymbol{\nabla} \cdot \mathbf{J}_2 = \sigma_2.$$

Therefore

$$\rho\mathrm{D}u - \overset{(i)}{\Big\{-p\rho\mathrm{D}\rho^{-1}} \overset{(ii)}{- T\boldsymbol{\nabla}\cdot(\mathbf{q}/T)} + \overset{(iii)}{[\rho T\mathrm{D}s_2 + T\boldsymbol{\nabla}\cdot\mathbf{J}_2]}\Big\} = T\sigma.$$

The labelled terms represent the power supplied reversibly to the thermodynamic system P by (i) compression, (ii) conduction and (iii) by relaxation of the internal structure. The stability of this process requires that the internal energy of the local thermodynamic system P should increase at a rate not less than that provided by the reversible power supply. Hence the left-hand side is non-negative, i.e.

$$\sigma = \sigma_1 + \sigma_2 \geq 0, \tag{41}$$

where (see [6])

$$\left.\begin{aligned}
\sigma_1 &= -T^{-1}(\boldsymbol{\pi}_1 : \overset{\circ}{\mathbf{e}} + \mathbf{q}_1 \cdot \boldsymbol{\nabla}\ln T), \\
\sigma_2 &= -T^{-1}(\boldsymbol{\pi}_2 : \overset{\circ}{\mathbf{e}} + \mathbf{q}_2 \cdot \boldsymbol{\nabla}\ln T) + \rho\mathrm{D}s_2 + \boldsymbol{\nabla}\cdot\mathbf{J}_2.
\end{aligned}\right\} \tag{42}$$

The functions required in (42) are collected here for the reader's convenience:

$$\left.\begin{aligned}
\mathbf{q}_1 &= -\kappa\boldsymbol{\nabla}T, \qquad \boldsymbol{\pi}_1 = -2\mu\overset{\circ}{\mathbf{e}}, \\
\boldsymbol{\pi}_2 &= -\tau_6\,\overset{\circ}{\overline{\mathbf{e}\cdot\boldsymbol{\pi}_1}} - \tau_7\,\overset{\circ}{\overline{\boldsymbol{\nabla}\mathbf{q}_1}} - \tau_8\boldsymbol{\pi}_1\boldsymbol{\nabla}\cdot\mathbf{v}, \\
\mathbf{q}_2 &= -\tau_3\mathrm{D}^*\mathbf{q}_1 - \tau_4\overset{\circ}{\mathbf{e}}\cdot\mathbf{q}_1 - \tau_5\mathbf{q}_1\boldsymbol{\nabla}\cdot\mathbf{v}, \\
\mathbf{J}_2 &= -\alpha\boldsymbol{\pi}_1\cdot\mathbf{q}_1 \qquad \alpha = (2/5pT), \\
s_2 &= -(4\rho pT)^{-1}(\boldsymbol{\pi}_1 : \boldsymbol{\pi}_1 + \tfrac{4}{5}\mathbf{q}_1\cdot\mathbf{q}_1/RT).
\end{aligned}\right\} \tag{43}$$
and

Kinetic theory [1] gives the following values for the relaxation times in (43):

$$\left.\begin{array}{llll}
\tau_3 = 2\tau_2, & \tau_4 = (\frac{14}{3} + \frac{8}{15}s)\tau_2, & \tau_5 = (3 + \frac{2}{3}s)\tau_2, \\
\tau_6 = 3\tau_1, & \tau_7 = \frac{4}{5}\tau_1, & \tau_8 = (3 - \frac{2}{3}s)\tau_1.
\end{array}\right\} \quad (44)$$

where
$$\tau_2 = \tfrac{3}{2}\tau_1 = \tfrac{3}{2}\mu/p = 2\kappa/(5Rp) > 0. \quad (45)$$

Two important features of the expressions for the fluxes $\mathbf{q}_2$, $\boldsymbol{\pi}_2$ and $\mathbf{J}_2$ are that they are (i) *reversible* and (ii) allow σ_2 to be of *either* sign. We define 'reversibility' in this context by requiring the constitutive terms to have the same parity under molecular motion reversal as the fluxes being defined. For example, the ideal parity η of $\mathbf{q}$ is -1, whereas that of $\mathbf{q}_1 \equiv -\kappa\nabla T$ is $+1$. Thus, since η is negative for $\nabla\mathbf{v}$, the second-order terms in

$$\overset{(i)}{}\quad\overset{(r)}{}\quad\overset{(r)}{}\quad\overset{(r)}{}$$
$$\mathbf{q} \approx \mathbf{q}_1 + \mathbf{q}_2 = -\kappa\nabla T - \tau_3 D^*\mathbf{q}_1 - \tau_4\overset{\circ}{\mathbf{e}}\cdot\mathbf{q}_1 - \tau_5\mathbf{q}_1\nabla\cdot\mathbf{v}$$

have the same parity as $\mathbf{q}$ and are therefore reversible, as indicated. Similarly, since the right-hand sides of $(42)_1$ and $(42)_2$ have values of η of $+1$ and -1 respectively,

$$\overset{(i)}{}\;\overset{(r)}{}$$
$$\sigma \approx \sigma_1 + \sigma_2,$$

and only σ_1 is dissipative. More generally, terms that are *even* powers of the microscopic times τ are reversible; dissipation is a property belonging to $\sigma_1, \sigma_3, \cdots$ and not to $\sigma_2, \sigma_4, \cdots$.

6.2. SIGN OF THE ENTROPY PRODUCTION RATE

As shown above, stability requires that

$$\sigma_1 \geq 0, \qquad \sigma_1 + \sigma_2 \geq 0. \quad (46)$$

It follows from $(42)_2$ and (43) that σ_2 may have either sign, which gives different rôles to the two inequalities in (46). The first determines that μ and κ in $(43)_1$ be positive numbers, in which sense it is a microscopic or *thermodynamic* constraint; it is universal and independent of the actual flows obtained. The coefficients of the second-order terms, being determined by the same molecular behaviour as the first-order coefficients, are closely related to them (see (44)), leaving no freedom for further thermodynamic constraints to be satisfied.

On the other hand the stability constraint $(\sigma_1 + \sigma_2) \geq 0$ can be satisfied only by restricting the *class* of fluid flows. It sets a limit on the gradients, or equivalently on the Knudsen number ε, making it a macroscopic or *fluid* constraint. Provided ε is sufficiently small—certainly less than unity—σ_1 will be larger than $|\sigma_2|$ and $(46)_2$ will be satisfied. We might imagine some initial state, with steep gradients, for which this does not hold. The fluid flow may be momentarily unstable, but the resulting fluctuation will quickly restore a 'new' equilibrium state, in which σ_2 is either positive, or smaller in magnitude than σ_1.

Since the second law of thermodynamics is not normally associated with the question of the stability or not of the flow field, we shall confine it to the thermodynamic constraint, $\sigma_1 \geq 0$ and regard the constraint on $(\sigma_1 + \sigma_2)$ as being concerned solely with the *macroscopic* stability of the flow field [7]. Certainly with magnetoplasmas, even with convergent Knudsen number expansions, the $O(\varepsilon^2)$ terms dominate the classical $O(\varepsilon)$ terms by orders of magnitude and instabilities in which $(\sigma_1 + \sigma_2)$ changes sign periodically do occur [6].

The fact that σ_2 may have both signs has induced some authors, concerned about a possible conflict with the second law of thermodynamics, to question the validity of the Knudsen number expansion. Eu [8] takes this view, which led him to develop his 'modified moment method', with the aim of having an entropy production rate in which *every* term is positive. However, the fact that it is possible to have velocity and temperature fields that render σ_2 negative, is no threat to the second law, at least as we have interpreted it above. A fluid instability requires σ_2 to be both negative and larger in magnitude than σ_1. This will occur only in exceptional circumstances. For small values of ε, it is easily confirmed from the above equations that, when σ_1 tends to zero as the gradients of the temperature and velocity become smaller and smaller, σ_2 also tends to zero, but at a greater rate. On the other hand if the gradients become very large, the Knudsen number expansion is not valid, and the difficulty is artificial.

6.3. CAN CONSTITUTIVE RELATIONS BE DEDUCED FROM $\sigma \geq 0$?

In the light of the classical derivation of the first-order constitutive relations from the thermodynamic inequality $\sigma_1 \geq 0$, the question arises 'Can anything more be learnt about constitutive relations from the fluid inequality $\sigma_1 + \sigma_2 \geq 0$?' To answer this, we shall first arrange $\sigma \approx \sigma_1 + \sigma_2$ into a bilinear form.

From (42) and $\boldsymbol{\pi} \approx \boldsymbol{\pi}_1 + \boldsymbol{\pi}_2$, $\mathbf{q} \approx \mathbf{q}_1 + \mathbf{q}_2$, we have

$$\sigma = -T^{-1}(\boldsymbol{\pi} : \overset{\circ}{\mathbf{e}} + \mathbf{q} \cdot \boldsymbol{\nabla}\ln T) + \rho \mathrm{D}s_2 + \boldsymbol{\nabla} \cdot \mathbf{J}_2,$$

where to sufficient accuracy

$$\mathrm{D}\ln\rho = -\boldsymbol{\nabla}\cdot\mathbf{v}, \quad \mathrm{D}\ln T = -\tfrac{2}{3}\boldsymbol{\nabla}\cdot\mathbf{v}, \quad \mathrm{D}\ln p = -\tfrac{5}{3}\boldsymbol{\nabla}\cdot\mathbf{v}. \tag{47}$$

Therefore from (10)

$$\rho D s_2 = -\left\{ \frac{5}{6pT} \boldsymbol{\pi}_1 : \boldsymbol{\pi}_1 + \frac{4\rho}{5p^2 T} \mathbf{q}_1 \cdot \mathbf{q}_1 \right\} \boldsymbol{\nabla} \cdot \mathbf{v}$$

$$- \frac{1}{2pT} \boldsymbol{\pi}_1 : D\boldsymbol{\pi}_1 - \frac{2\rho}{5p^2 T} \mathbf{q}_1 \cdot D\mathbf{q}_1.$$

Also from $(43)_5$,

$$\boldsymbol{\nabla} \cdot \mathbf{J}_2 = -\boldsymbol{\nabla} \cdot (\alpha \boldsymbol{\pi}_1 \cdot \mathbf{q}_1) \qquad (\alpha \equiv 2/(5pT)). \qquad (48)$$

In the second-order expressions we may replace $\boldsymbol{\pi}_1$ and $\mathbf{q}_1$ by $\boldsymbol{\pi}$ and $\mathbf{q}$ without changing the order of the error term. Making these changes and using (45), we can write σ in the form

$$\sigma = (\kappa T^2)^{-1} \mathbf{q} \cdot \left\{ -\kappa \boldsymbol{\nabla} T - \tau_2 D^* \mathbf{q} - \kappa T^2 \boldsymbol{\nabla} \cdot (\alpha \boldsymbol{\pi}) - 2\tau_2 \mathbf{q} \boldsymbol{\nabla} \cdot \mathbf{v} \right\}$$

$$+ (2\mu T)^{-1} \boldsymbol{\pi} : \left\{ -2\mu \overset{\circ}{\mathbf{e}} - \tau_1 D^* \boldsymbol{\pi} - 2\mu T \alpha \boldsymbol{\nabla} \mathbf{q} - \tfrac{5}{3} \tau_1 \boldsymbol{\pi} \boldsymbol{\nabla} \cdot \mathbf{v} \right\}. \qquad (49)$$

Here we have used $\mathbf{q} \cdot D^* \mathbf{q} = \mathbf{q} \cdot (D\mathbf{q} - \boldsymbol{\Omega} \times \mathbf{q}) = \mathbf{q} \cdot D\mathbf{q}$ and similarly $\boldsymbol{\pi} : D^* \boldsymbol{\pi} = \boldsymbol{\pi} : D\boldsymbol{\pi}$, to introduce frame-indifferent derivatives into σ. Notice that we have now obliterated any trace of the order of the terms. Equation (49) is accurate up to terms $O(\varepsilon^2)$. To this order it is quite general, no commitment having been made about the actual values of $\mathbf{q}$ and $\boldsymbol{\pi}$—for example there has been no linearization. Nor are we free to adjust the coefficients: τ_1 and τ_2 are defined in (45) and α is chosen to given agreement with kinetic theory.

Stability ('thermodynamic' plus 'fluid') requires that $\sigma \geq 0$, but treating this as being entirely a *thermodynamic* constraint, we may satisfy it by setting the first bracket in (49) equal to $\mathbf{q}$ and the second equal to $\boldsymbol{\pi}$. With this choice, we get

$$\left. \begin{aligned} \mathbf{q} &= -\kappa \boldsymbol{\nabla} T - \tau_2 D^* \mathbf{q} - \kappa T^2 \boldsymbol{\nabla} \cdot (\alpha \boldsymbol{\pi}) - 2\tau_2 \mathbf{q} \boldsymbol{\nabla} \cdot \mathbf{v}, \\ \boldsymbol{\pi} &= -2\mu \overset{\circ}{\mathbf{e}} - \tau_1 D^* \boldsymbol{\pi} - \tfrac{4}{5} \tau_1 \boldsymbol{\nabla} \mathbf{q} - \tfrac{5}{3} \tau_1 \boldsymbol{\pi} \boldsymbol{\nabla} \cdot \mathbf{v}. \end{aligned} \right\} \qquad (50)$$

But these equations are not correct. The most obvious error is the omission of the non-linear term in $\boldsymbol{\pi}_2$—the first right-hand term in $(43)_3$. This term arises, not only in the theory developed in §4.1, but also in the kinetic theory based on Boltzmann's equation. Also the non-linear 'τ_4' term appearing in $(43)_4$ is absent from $\mathbf{q}$, although a similar term can be deduced from $-\kappa T^2 \boldsymbol{\nabla} \cdot (\alpha \boldsymbol{\pi})$. This may be written

$$-\kappa T^2 \boldsymbol{\nabla} \cdot (\alpha \boldsymbol{\pi}) = -\tau_2 RT \boldsymbol{\nabla} \cdot \boldsymbol{\pi} - 3\tau_1^2 RT \boldsymbol{\nabla} p \cdot \overset{\circ}{\mathbf{e}} + \tfrac{4}{5} \tau_1 \overset{\circ}{\mathbf{e}} \cdot \mathbf{q}_1,$$

but this has the wrong sign. And for the reasons given in §3.1, the other two terms should be absent. These latter terms *do* appear in the Burnett equations, but this is due to a defect in Boltzmann's integro-differential equation for f ([1], p. 160).

Our conclusion is that the fluid constraint, $(\sigma_1 + \sigma_2) \geq 0$ can tell us nothing about the constitutive equations themselves. Its rôle is to identify a class of unstable fluid flows. An important example of an unstable flow of this type is described in [6].

7. Extended Irreversible Thermodynamics

7.1. BASIC CONCEPTS

In recent years new approaches to the problem of formulating constitutive equations, in conditions well removed from equilibrium, have been developed by a number of scientists. The most important and useful of these is described as 'extended irreversible thermodynamics'. The principal contributors are Nettleton [9], Müller [10], Garcia–Colin [11, 12], Eu [13] and Jou, Casas–Vàzquez and Lebon [14, 15]. The last named authors have recently published a text [3] explaining the subject and applying it to a number of important problems. In a brief survey [16] they state:

> This new approach ... was born out of the double necessity to go beyond the hypothesis of local equilibrium and to avoid the paradox of propagation of disturbances with infinite velocity. This physically unpleasant property arises as a consequence of using the Fourier, Newton and Fick laws in the balance equations of energy, momentum and mass, respectively. Indeed the resulting equations are parabolic from which it follows that disturbances will be felt instantaneously in the whole of space.

'Local equilibrium', which is not a precise concept is not as restrictive as the quotation implies. It certainly does not rule out second-order transport. But if 'going beyond the hypothesis of local equilibrium' means ignoring Knudsen number constraints, then collective or 'fluid' behaviour would not be well represented. That the 'paradox' referred to was ill-conceived (see §5.2) does not render the new approach illegitimate.

The innovations in extended irreversible thermodynamics (EIT) are (i) to include the dissipative fluxes $\mathbf{q}$ and $\boldsymbol{\pi}$ among the set of variables defining state space, (ii) to specify evolution equations for these fluxes and (iii) to choose these constitutive equations so that the rate of entropy production, σ, is always positive. Knudsen number expansions are not part of the theory, although the importance of not allowing ε to become too large is recognized—if not always respected. The distinction between first and second-order terms, is not required which means that the inequality $\sigma \geq 0$ is always treated as being a thermodynamic constraint on the constitutive equations.

A central rôle is assigned to a generalized entropy $s(u, \rho, \mathbf{q}, \mathbf{\pi}, \overset{\times}{\pi})$, which is assumed to be a given function. In many, if not most of the applications, s is taken to be a quadratic function of the new arguments $\mathbf{q}$ and $\mathbf{\pi}$, an approximation that in effect limits EIT to the same range as the second-order theory described in earlier sections of this article.

7.2. KINETIC THEORY PRECURSORS OF EIT

Three important early contributions to higher-order transport, that have had an influence on EIT, should be noted here.

Equation (122) of Maxwell's [4] seminal paper, *The Dynamical Theory of Gases*, was probably the first constitutive relation to include time derivatives. In our notation it reads

$$\frac{\partial \pi_{xx}}{\partial t} + 2p\frac{\partial v_x}{\partial x} - \tfrac{2}{3}p\nabla \cdot \mathbf{v} = \frac{\delta \pi_{xx}}{\delta t}, \qquad (51)$$

where the right-hand term is the rate of change of the specified component of the viscous stress tensor due to molecular collisions. Maxwell showed that this could be expressed as $-\pi_{xx}/\tau_1$, and for the case when π_{xx} is not maintained by the motion of the medium, he deduced the law $\pi_{xx} \propto \exp(-t/\tau_1)$. He therefore described τ_1 as being 'the modulus of the time of relaxation'. Equation (51) is one component of the general relation

$$\tau_1 D\mathbf{\pi} + \mathbf{\pi} = -2\mu\overset{\circ}{\mathbf{e}}. \qquad (52)$$

Maxwell well understood the limitations of including a time-derivative:

If the motion is not subject to any very rapid changes, as in all cases except that of the propagation of sound, we may neglect $\partial \pi_{xx}/\partial t$.

In a later paper, Maxwell [17] also obtained the linear component of the term in $\mathbf{\pi}$ containing $\nabla \mathbf{q}$ (see (43)$_3$), namely that involving $\nabla\nabla T$. He developed this in order to provide an explanation of the rotation of the vanes of a radiometer (see [18, pp. 210–219).

Burnett [19] obtained a complete set of second-order transport terms by applying an expansion in powers of ε to Boltzmann's equation. Other notable contributions to the theory were made by Chapman and Cowling [20] and Grad [21, 22]. The velocity distribution function f is expanded in the form,

$$f = f_0\{1 + \varphi_1 + \varphi_2 + \cdots\} \qquad (\varphi_r = O(\varepsilon^r)), \qquad (53)$$

where f_0 is the equilibrium distribution. To a good approximation, it can be shown that (e.g. see [1])

$$\varphi_1 = -\tau_2(\nu^2 - \tfrac{5}{2})\mathbf{c} \cdot \boldsymbol{\nabla} \ln T - \tau_1 \boldsymbol{\nu} \cdot \overset{\circ}{\mathbf{e}} \cdot \boldsymbol{\nu}, \tag{54}$$

where $\boldsymbol{\nu} = \mathbf{c}/C$, $C \equiv (2k_B T/m)^{\frac{1}{2}}$ and $\mathbf{c}$ is the molecular velocity measured in the local thermodynamic system P. Burnett's contribution was to determine the value of φ_2 and to use this to calculate expressions for $\mathbf{q}_2$ and $\boldsymbol{\pi}_2$.

The expressions he obtained (see [23], chapter 15 or [1], chapter 8) for $\mathbf{q}_2$ and $\boldsymbol{\pi}_2$ are

$$\mathbf{q}_2 = Rp\tau_1^2\Big\{\theta_1 \boldsymbol{\nabla} \cdot \mathbf{v}\, \boldsymbol{\nabla} T + \theta_2(D\boldsymbol{\nabla} T - \boldsymbol{\nabla}\mathbf{v} \cdot \boldsymbol{\nabla} T) + \theta_3 Tp^{-1}\boldsymbol{\nabla} p \cdot \overset{\circ}{\mathbf{e}}$$

$$+\theta_4 T\boldsymbol{\nabla} \cdot \overset{\circ}{\mathbf{e}} + 3\theta_5 \boldsymbol{\nabla} T \cdot \overset{\circ}{\mathbf{e}}\Big\}, \tag{55}$$

with $\quad \theta_1 = \tfrac{15}{4}(\tfrac{7}{2} - \mathsf{s}), \quad \theta_2 = \tfrac{45}{8}, \quad \theta_3 = -3, \quad \theta_4 = 3, \quad \theta_5 = \tfrac{35}{4} + \mathsf{s} \tag{56}$

and $\quad \boldsymbol{\pi}_2 = p\tau_1^2\Big\{\tilde{\omega}_1 \boldsymbol{\nabla} \cdot \mathbf{v}\,\overset{\circ}{\mathbf{e}} + \tilde{\omega}_2(D\overset{\circ}{\mathbf{e}} - 2\,\overline{\boldsymbol{\nabla}\mathbf{v} \cdot \overset{\circ}{\mathbf{e}}}) + \tilde{\omega}_3 R\,\overline{\boldsymbol{\nabla}\boldsymbol{\nabla} T}$

$$+\tilde{\omega}_4 Rp^{-1}\overline{\boldsymbol{\nabla} p\boldsymbol{\nabla} T} + \tilde{\omega}_5 RT^{-1}\overline{\boldsymbol{\nabla} T\boldsymbol{\nabla} T} + \tilde{\omega}_6\,\overline{\overset{\circ}{\mathbf{e}} \cdot \overset{\circ}{\mathbf{e}}}\Big\}, \tag{57}$$

with $\quad \tilde{\omega}_1 = \tfrac{4}{3}(\tfrac{7}{2} - \mathsf{s}), \quad \tilde{\omega}_2 = 2, \quad \tilde{\omega}_3 = 3, \quad \tilde{\omega}_4 = 0, \quad \tilde{\omega}_5 = 3\mathsf{s}, \quad \tilde{\omega}_6 = 8. \tag{58}$

These equations have been used by those developing EIT to justify their constitutive equations and to find values for some of their phenomenological parameters, at least for the special case of a tenuous gas. Jou, Casas–Vàzquez and Lebon ([3], p. 81) state:

> The agreement between the kinetic theory of gases and the macroscopic predictions [of EIT theory] undoubtably reinforces the consistency of extended irreversible thermodynamics. Although our anaylsis is limited to the linear range, the accord between EIT and kinetic theory is much wider than the agreement of the latter with the usual local equilibrium thermodynamics.

Grad's [21, 22] method (in a simplified form) was to replace the gradients in (54) by $\mathbf{q}_1$ and $\boldsymbol{\pi}_1$ and then to generalize the result by substituting $\mathbf{q}$ and $\boldsymbol{\pi}$ in their place. This amounts to replacing $(\varphi_1 + \varphi_2 + \cdots)$ in (53) by the single term

$$\varphi = \frac{2\rho}{5p^2}(\nu^2 - \tfrac{5}{2})\mathbf{c} \cdot \mathbf{q} + \frac{\rho}{2p^2}\,\overset{\circ}{\mathbf{cc}} : \boldsymbol{\pi}. \tag{59}$$

The time derivatives of $\mathbf{q}$ and $\boldsymbol{\pi}$ are now obtained by substituting the function $f = f_0(1 + \varphi)$ into the appropriate moments of Boltzmann's equation.

Grad's approach to higher-order transport was particularly relevant to EIT, for by expressing the distribution function in terms of *fluxes* rather than gradients, he initiated a new way of treating kinetic theory.

7.3. CONSTITUTIVE EQUATIONS FOR A ONE-COMPONENT FLUID

We shall sketch the EIT approach to deriving constitutive relations for a one-component gas, following the account given in chapter 2 of Jou, Casas–Vàzquez and Lebon [3]. The independent variables are taken to be u, ρ, $\mathbf{q}$, and $\boldsymbol{\pi}$; the trace of $\boldsymbol{\pi}$ has been omitted to slightly simplify the equations. The authors begin by supposing $s = s(u, \rho, \mathbf{q}, \boldsymbol{\pi})$ to be a known function, with s quadratic in the fluxes $\mathbf{q}$ and $\boldsymbol{\pi}$ (cf. (10)). Similarly the entropy flux vector is postulated to have the form

$$\mathbf{J} = \mathbf{q}/T - \alpha(u, \rho)\boldsymbol{\pi} \cdot \mathbf{q}, \tag{60}$$

where the leading term recovers the classical (first-order) theory and the second term is the most general (second-order) expression involving $\mathbf{q}$ and $\boldsymbol{\pi}$.

The entropy production rate now follows directly from

$$\sigma = \rho D s + \boldsymbol{\nabla} \cdot \mathbf{J} \geq 0, \tag{61}$$

it being assumed that σ is always positive. In our notation they find that the entropy production rate is given by

$$\sigma = \ (\kappa T^2)^{-1}\mathbf{q} \cdot \{-\kappa\boldsymbol{\nabla}T - \tau_2 D\mathbf{q} - \kappa T^2\boldsymbol{\nabla} \cdot (\alpha\boldsymbol{\pi})\}$$
$$+ (2\mu T)^{-1}\boldsymbol{\pi} : \{-2\mu\overset{\circ}{\mathbf{e}} - \tau_1 D\boldsymbol{\pi} - 2\mu T\alpha \overset{\circ}{\boldsymbol{\nabla}\mathbf{q}}\}. \tag{62}$$

Except for lacking the terms in $\boldsymbol{\nabla} \cdot \mathbf{v}$, this expression agrees with (49). At this stage the authors observe that the entropy production rate is independent of whether one adopts frame-indifferent derivatives or not. They come to no definite conclusion on the question of which derivative to use, noting that in kinetic theory frame-dependent derivatives do occur[2].

The method is certainly simpler than that of §2 and 3, where we were concerned with the *order* of the terms, but it has the disadvantage of being phenonmenological and formal rather than physical. Perhaps this is a matter of taste; what finally matters is whether or not the theory correctly *represents* the physics.

---

[2]These are easily avoided by defining the molecular velocity in a rotating frame [24] .

The next step is to secure the entropy inequality by adopting a linear relation between the terms in the brackets and the fluxes **q** and **π**. This yields constitutive equations, that excepting the $\nabla \cdot \mathbf{v}$ term, are identical with (50). And they suffer from the same defects. They cannot be said to be in agreement with the results of kinetic theory, unless restricted to their linear terms.

We are discussing only the foundations of EIT. In its many successful and wide-ranging applications, just how the constitutive equations are justified formally is not paramount. The successes follow naturally when the linear higher-order terms (generally second-order in the Knudsen number) become important.

References

[1] Woods, L.C. (1993). *Kinetic theory of gases and magnetoplasmas*. Oxford University Press.

[2] Woods, L.C. (1986). *The thermodynamics of fluid systems*. Oxford University Press.

[3] Jou, D., Casas–Vázquez, J. and Lebon, G. (1993). *Extended irreversible thermodynamics*. Springer–Verlag.

[4] Maxwell, J.C. (1867). *Phil. Trans. Roy. Soc.* **157**, 49–88; (See also *Scientific papers of James Clerk Maxwell*, W.D. Niven, ed. New York: Dover, (1965), vol. 2, 43.)

[5] Cattaneo, C. (1948). *Atti Sem. Mat. Fis. Univ. Modena.* **3**, 83.

[6] Woods, L.C. (1994). *J. Non-equilib. Thermodynam.* **19**, 61–75.

[7] Woods, L.C. (1982). *IMA Journ. Appl. Maths.* **29**, 221–46.

[8] Eu, B.C. (1992). *Kinetic theory and irreversible thermodynamics*. John Wiley & Sons, Inc.

[9] Nettleton, R. (1960). *Phys. Fluids*, **3**, 216.

[10] Müller, I. (1967). *Z. Physik.* **198**, 329.

[11] Garcia–Colin, L., Lopez de Haro, M., Rodriquez, R., Casas–Vázquez, J. and Jou, D. (1984). *J. Stat. Phys.* **37**, 465.

[12] Garcia–Colin, L. (1988). *Rev. Mex. Fisica.* **34**, 344.

[13] Eu, B.C. (1980). *J. Chem. Phys.* **73**, 2958.

[14] Jou, D., Casas–Vázquez, J. and Lebon, G. (1979). *J. Non-equilib. Thermodynam.* **4**, 31.

[15] Lebon, G., Casas–Vázquez, J. and Jou, D. (1980). *J. Phys.* A, **13**, 275.

[16] Lebon, G., Jou, D., and Casas–Vázquez, J. (1992). *Contemp. Phys.* **33**, (1), 41–51.

[17] Maxwell, J.C. (1879). *Phil. Trans. Roy. Soc.* **170**, 231–56.

[18] Brush, S.G. (1976). *The kind of motion we call heat.* 2 Vols. North-Holland Pub. Co., Amsterdam.

[19] Burnett, D. (1935). *Proc. London Math. Soc.* **39**, 385–430, and in **40**, 382—435.
[20] Chapman, S. and Cowling, T.G. (1941). *Proc. Roy. Soc. A.* **179**, 159.
[21] Grad, H. (1949). *Comm. Pure and Appl. Math.* **2**, 311–407.
[22] Grad, H. (1952). *Comm. Pure and Appl. Math.* **5**, 257.
[23] Chapman, S. and Cowling, T.G. (1970). *The mathematical theory of non-uniform gases*. Cambridge University Press.
[24] Woods, L.C. (1983). *J. Fluid Mech.* **136**, 423–433.

MESOSCOPIC THEORY OF LIQUID CRYSTALS

W. MUSCHIK, H. EHRENTRAUT, C. PAPENFUß AND S. BLENK

Institut für Theoretische Physik
Technische Universität Berlin
Hardenbergstraße 36, 10623 Berlin, Germany

Abstract. The structural roots of the mesoscopic theory of liquid crystals are reviewed and shortly discussed.

1. Introduction

Liquid crystals are phases showing an orientational order of molecules which are of elongated or of plane shape, so that an orientational order of them can be defined. The molecules themselves may be uniaxial (i.e. their alignment can be described by one direction) or biaxial (for which two axes are necessary to fix their orientation). Besides the orientational order an additional spatial order of the centers of mass of the molecules is possible which causes a great variety of liquid crystal phases, such as nematics, smectics, cholesterics and a lot of other phases of various different structures. Because of thermal fluctuations the molecules are not totally aligned, but they have a certain distribution around a "mean orientation" which can be described by a normalized *macroscopic director field.* The name "macroscopic" originates from the fact, that the macroscopic director field belongs to all molecules of a volume element, whereas a special molecule may be aligned differently. So it is obvious to introduce a *microscopic director* which describes the alignment of a single molecule and which is different from the local macroscopic director (if it exists). Because the microscopic director is defined on a molecular level it is not a macroscopic field, but a so-called *mesoscopic variable.* Here "mesoscopic" means that the level of description is finer than the macroscopic level is, but that no microscopic concept such as molecular interaction or potentials are used.

J. S. Shiner (ed.), Entropy and Entropy Generation, 101–109.

Besides the microscopic and the macroscopic director other descriptions for alignment are in use which can be found in the following synopsis:

Up to now there are five different phenomenological concepts suitable to describe liquid crystals: The first one is the well known Ericksen-Leslie theory [1, 2] whose balance equations are formulated by use of the macroscopic director field mentioned above. But in fact this theory is not able to represent a change in the degree of orientational order. In general we need at each position and time a distribution function for describing the macroscopic orientation of the fluid. Therefore the macroscopic director has to be redefined statistically.

The second concept describes liquid crystals as micropolar media in the frame of a 3-director theory [3]. Instead of a balance equation for the macroscopic director the spin balance is taken into account, but no microscopic concepts are used.

The third concept [4] introduces besides the balance equations of a micropolar media an additional field, called microinertia tensor field. This field satisfying its own balance equation is coupled to the spin balance. The form of this coupling causes that all needle-shaped molecules of a volume element have always the same angular velocity. Therefore the degree of alignment cannot change (in a co-moving frame) as it happens in the phase transition from nematic to isotropic phase. Consequently the microinertia tensor field – although the molecules are not totally aligned – is not appropriate to describe dynamical situations in which the degree of alignment changes.

The fourth concept describing liquid crystals uses the alignment tensor [5, 6], a symmetric, traceless tensor which represents the first anisotropy moment of a multipole expansion of the orientation distribution function. This of course is a more general method for describing alignment as using a director theory because the alignment tensor is determined by five, the director by two variables.

The fifth concept describing liquid crystals introduces the above mentioned microscopic director related to the orientation of one rigid particle only. The anisotropic fluid is formally treated as a mixture by regarding all particles of a volume element of the same orientation as one component of the fluid [7]. Thus the orientation distribution function results as the fraction of the mass density of one component over the mass density of the mixture. Because mixture theories are well developed [8, 9] balance equations for liquid crystals can be written down very easily by use of this method. The domain on which these balances are defined is different from the usual one because it contains the microscopic director as an additional variable.

The aim of this paper is to sketch the mesoscopic theory of liquid crystals which is based on the concepts of the microscopic director and of the orientation distribution function.

2. Mesoscopic Concept

As discussed above a liquid crystal is composed of differently orientated particles for which we presuppose, that their orientation can be described individually by the direction of their microscopic directors. Thus the microscopic director n is defined as a unit vector pointing into the temporary direction of a needle-shaped rigid particle, or, if the particle is of a plane shape, the microscopic director is perpendicular to the particle. Because the microscopic director is normalized

$$n^2 = 1, \quad n \in S^2, \tag{1}$$

the microscopic alignment has two degrees of freedom. Since the particles may rotate, the microscopic director changes in time and we define the microscopic orientation change velocity u by

$$u := \frac{d}{dt}n, \quad \text{with} \quad u \cdot n = 0. \tag{2}$$

If we consider molecules of a volume element the alignment of their axes is specified by a distribution function f on the unit sphere, called *orientation distribution function* (ODF). The ODF describes the density generated by intersection points on the unit sphere S^2 between the molecule axes and the sphere. Because in general the alignment of the molecules is a function of position and time the ODF is defined on a six-dimensional space $S^2 \times \mathbb{R}^3 \times \mathbb{R}^1$

$$f(n, x, t) \equiv f(\cdot), \quad (\cdot) \equiv (n, x, t) \in S^2 \times \mathbb{R}^3 \times \mathbb{R}^1. \tag{3}$$

The 5-dimensional subspace $S^2 \times \mathbb{R}^3$ consisting of the orientation- and of the position-part is called *nematic space*. Because there are always two points of intersection opposite to each other on the S^2 the ODF shows the so-called head-tail symmetry

$$f(-n, x, t) = f(n, x, t). \tag{4}$$

By this ODF a mesoscopic classification of different types of orientation, so-called *phases*, becomes possible.

3. Orientational Balances

In the nematic space Ω each particle is represented by its five coordinates (x, n) at time t. We define fields on the nematic space, such as the field of orientational mass density which is defined by

$$\rho(\cdot) := \rho(x, t)f(\cdot), \tag{5}$$

where $\rho(\boldsymbol{x}, t)$ is the macroscopic mass density (25) of the considered one-component liquid crystal. $\rho(\cdot)$ describes the mass density of those molecules having the orientation $\boldsymbol{n}$ at position $\boldsymbol{x}$ at time t. In this interpretation $\boldsymbol{v}(\cdot)$ is the material velocity of these molecules. Other such orientational fields (i.e. fields defined on the nematic space) are e.g. the external acceleration $\boldsymbol{k}(\cdot)$, the stress tensor $\boldsymbol{T}(\cdot)$, and the heat flux density $\boldsymbol{q}(\cdot)$.

3.1. GLOBAL BALANCES

Using these orientational fields global balances on the nematic space can be formulated [10] which all have the following shape

$$\frac{\mathrm{d}}{\mathrm{d}t} \int_{G_t} \rho(\cdot)[\phi(\cdot)\boldsymbol{v}(\cdot) + \boldsymbol{\Phi}(\cdot)] \, d^2n d^3x = \boldsymbol{Z}^{tot}, \tag{6}$$

$$\boldsymbol{Z}^{tot} = \int_{G_t} [\rho(\cdot)\boldsymbol{K}(\cdot) + \nabla_x \cdot \boldsymbol{S}(\cdot)] \, d^2n d^3x. \tag{7}$$

Here $\phi(\cdot)$ is the convective part of the balance, $\boldsymbol{K}(\cdot)$ the volume part, and $\boldsymbol{S}(\cdot)$ the surface part of the external quantity $\boldsymbol{Z}^{tot}$.

We obtain for the special balances the following identities [11]:

Mass

$$\phi(\cdot) \equiv 0, \quad \boldsymbol{\Phi}(\cdot) \ \equiv \ 1, \quad \boldsymbol{Z}^{tot} \equiv \boldsymbol{0}, \tag{8}$$

$$\boldsymbol{K}(\cdot) \ \equiv \ 0, \quad \boldsymbol{S}(\cdot) \equiv \boldsymbol{0}, \tag{9}$$

Momentum

$$\phi(\cdot) \equiv 1, \ \boldsymbol{\Phi}(\cdot) \ \equiv \ \boldsymbol{0}, \ \boldsymbol{Z}^{tot} = \text{total external force}, \tag{10}$$

$$\boldsymbol{K}(\cdot) \ \equiv \ \boldsymbol{k}(\cdot) = \text{external acceleration}, \tag{11}$$

$$\boldsymbol{S}(\cdot) \ \equiv \ \boldsymbol{T}^{\mathsf{T}}(\cdot) = \text{transposed stress tensor}, \tag{12}$$

Angular Momentum

$$\phi(\cdot) \equiv \boldsymbol{x} \times, \boldsymbol{\Phi}(\cdot) \ \equiv \ I\boldsymbol{n} \times \boldsymbol{u}(\cdot), \boldsymbol{Z}^{tot} = \text{total external moment}, \tag{13}$$

$$\boldsymbol{K}(\cdot) \ \equiv \ \boldsymbol{x} \times \boldsymbol{k}(\cdot) + \boldsymbol{n} \times \boldsymbol{g}(\cdot), \tag{14}$$

$$\boldsymbol{S}(\cdot) \ \equiv \ [\boldsymbol{x} \times \boldsymbol{T}(\cdot)]^{\mathsf{T}} + \boldsymbol{n} \times \boldsymbol{\pi}(\cdot), \tag{15}$$

Total Energy

$$\phi(\cdot) \equiv (1/2)\boldsymbol{v}(\cdot), \boldsymbol{\Phi}(\cdot) \ \equiv \ (1/2)I[\boldsymbol{n} \times \boldsymbol{u}(\cdot)]^2 + \varepsilon(\cdot), \tag{16}$$

$$\boldsymbol{Z}^{tot} \ = \ \text{energy supply}, \tag{17}$$

$$\boldsymbol{K}(\cdot) \ \equiv \ \boldsymbol{k}(\cdot) \cdot \boldsymbol{v}(\cdot) + \boldsymbol{g}(\cdot) \cdot \boldsymbol{u}(\cdot) + r(\cdot), \tag{18}$$

$$\boldsymbol{S}(\cdot) \ \equiv \ \boldsymbol{v}(\cdot) \cdot \boldsymbol{T}(\cdot) + \boldsymbol{u}(\cdot) \cdot \boldsymbol{\pi}(\cdot) - \boldsymbol{q}(\cdot). \tag{19}$$

Here $u(\cdot)$ is the field of the orientational change velocity, $g(\cdot)$ the orientational couple force density, $\pi(\cdot)$ the orientational couple stress tensor, $\varepsilon(\cdot)$ the density of orientational internal energy, and $r(\cdot)$ the orientational radiation supply.

Having taken into account the balances of mass, momentum, angular momentum, and energy we got all balances which are relevant for liquid crystals outside electrodynamics. These global balance equations formulated on the nematic space can be transformed into so-called orientational balance equations which are local in position, time, and the microscopic director. For this purpose we need a generalized Reynolds' transport theorem [10].

$$\frac{\mathrm{d}}{\mathrm{d}t} \int_{G_t} \Phi(\cdot) d^2n d^3x =$$

$$= \int_{G_t} \left[\frac{\partial}{\partial t} \Phi(\cdot) + \nabla_x \cdot [v(\cdot)\Phi(\cdot)] + \nabla_n \cdot [u(\cdot)\Phi(\cdot)] \right] d^2n \, d^3x. \quad (20)$$

By use of (20) we can transform the global balance equations (6) as usual into local ones.

3.2. LOCAL ORIENTATIONAL BALANCES

Presupposing the global balances we obtain [12] by use of (20) local balance equations which are defined on the nematic space and which are therefore denoted as *local orientational balances*. The general shape of these balances is

$$\frac{\partial}{\partial t}\{\rho(\cdot)[\phi(\cdot)v(\cdot) + \Phi(\cdot)]\} + \nabla_x \cdot \{\rho(\cdot)v(\cdot)[\phi(\cdot)v(\cdot) + \Phi(\cdot)]\} +$$

$$+\nabla_n \cdot \{\rho(\cdot)u(\cdot)[\phi(\cdot)v(\cdot) + \Phi(\cdot)]\} = \rho(\cdot)K(\cdot) + \nabla_x \cdot S(\cdot). \quad (21)$$

Besides the local balances of mass, momentum, angular momentum, and energy we obtain from the balance of momentum by subtracting the balance of angular momentum the *orientational spin balance*:

$$s(\cdot) := In \times u(\cdot) = \text{orientational spin density}, \quad (22)$$

$$\frac{\partial}{\partial t}[\rho(\cdot)s(\cdot)] + \nabla_x \cdot [\rho(\cdot)v(\cdot)s(\cdot) - (n \times \pi(\cdot))^T] +$$

$$+\nabla_n \cdot [\rho(\cdot)u(\cdot)s(\cdot)] = \varepsilon : T(\cdot) + \rho(\cdot)n \times g(\cdot) \quad (23)$$

(ε is the Levi-Civita tensor). The right-hand side of this equation indicates that one part of the spin supply is caused by the antisymmetric part of the stress tensor.

According to the definition of the orientational mass density (5) we obtain from the orientational mass balance an additional balance of the ODF by inserting the definition of $f(\cdot)$:

$$\frac{\partial}{\partial t} f(\cdot) + \nabla_x \cdot (v(\cdot) f(\cdot)) + n \cdot \nabla_n \times (\omega(\cdot) f(\cdot)) +$$

$$+ f(\cdot)[\frac{\partial}{\partial t} + v(\cdot) \cdot \nabla_x] \ln \rho(x, t) = 0. \tag{24}$$

Because this balance equation of the orientation distribution function includes the macroscopic field of the mass density

$$\rho(x, t) := \int_{S^2} \rho(\cdot) d^2 n , \tag{25}$$

it is not independent of the macroscopic mass balance defined on $\mathbf{R}^3$. As can be seen from (25) macroscopic quantities are defined by integration over the orientational part.

3.3. BALANCES OF MICROPOLAR MEDIA

Introducing the macroscopic fields by mesoscopic definitions the orientational balance equations transform into the balances for micropolar media by integrating over the unit sphere [10]. So the definition of the macroscopic material velocity

$$\rho(x, t) v(x, t) := \int_{S^2} \rho(\cdot) v(\cdot) d^2 n , \tag{26}$$

and (25) results in the macroscopic mass balance. The shape of the macroscopic balances by integrating (21) is

$$\frac{\partial}{\partial t} \{\rho(x, t)[\phi(x, t) v(x, t) + \Phi(x, t)]\} + \tag{27}$$

$$+ \nabla_x \cdot \{\rho(x, t) v(x, t)[\phi(x, t) v(x, t) + \Phi(x, t)]\} =$$

$$= \rho(x, t) K(x, t) + \nabla_x \cdot S(x, t). \tag{28}$$

Here the following mesoscopic definitions of macroscopic quantities are introduced:

$$\rho(x, t) \phi(x, t) := \int_{S^2} \rho(\cdot) \phi(\cdot) d^2 n , \tag{29}$$

$$\rho(x, t) v(x, t) := \int_{S^2} \rho(\cdot) v(\cdot) d^2 n , \tag{30}$$

$$\rho(x, t) K(x, t) := \int_{S^2} \rho(\cdot) K(\cdot) d^2 n , \tag{31}$$

$$\rho(\boldsymbol{x}, t)\boldsymbol{\Phi}(\boldsymbol{x}, t) := \int_{S^2} \rho(\cdot)[\phi(\cdot)\boldsymbol{v}(\cdot) + \boldsymbol{\Phi}(\cdot)]d^2n +$$
$$+ \rho(\boldsymbol{x}, t)\phi(\boldsymbol{x}, t)\boldsymbol{v}(\boldsymbol{x}, t), \tag{32}$$

$$S(\boldsymbol{x}, t) := \int_{S^2} \{S(\cdot) - \rho(\cdot)\boldsymbol{v}(\cdot)[\phi(\cdot)\boldsymbol{v}(\cdot) + \boldsymbol{\Phi}(\cdot)]\}d^2n -$$
$$- \rho(\boldsymbol{x}, t)\boldsymbol{v}(\boldsymbol{x}, t)[\phi(\boldsymbol{x}, t)\boldsymbol{v}(\boldsymbol{x}, t) + \boldsymbol{\Phi}(\boldsymbol{x}, t)]. \tag{33}$$

If the ϕ, $\boldsymbol{\Phi}$, $\boldsymbol{K}$, and S are specified according to (8)-(19) we obtain the macroscopic balances of mass, momentum, spin, and energy. Besides these balances two more ones are essential: the entropy and the alignment tensor balance. The entropy balance reads

$$\frac{\partial}{\partial t}[\rho(\boldsymbol{x}, t)\eta(\boldsymbol{x}, t)] + \nabla_x \cdot [\rho(\boldsymbol{x}, t)\eta(\boldsymbol{x}, t)\boldsymbol{v}(\boldsymbol{x}, t) - \phi(\boldsymbol{x}, t)] =$$
$$= \rho(\boldsymbol{x}, t)\sigma(\boldsymbol{x}, t) \tag{34}$$

(η = field of specific entropy, ϕ = entropy flux density, σ = entropy production density). The second law is expressed by

$$\sigma(\boldsymbol{x}, t) \geq 0. \tag{35}$$

The alignment tensor balance is sketched in the next section.

4. Alignment Tensors

The mesoscopic description of liquid crystals offers the advantage to introduce macroscopic fields, denoted as alignment tensor fields, which describe orientation more precise than the field of the macroscopic director. The main idea of this approach is expand the ODF $f(\cdot)$ into a series of multipoles [10, 13]. The coefficients of that expansion are tensors of even order. The balance equation (24) of the ODF induces alignment tensor balance equations of the following structure

$$\left(\frac{\partial}{\partial t} + \boldsymbol{v}(\boldsymbol{x}, t) \cdot \nabla_x\right) a_{\nu_1 \cdots \nu_k} - k\left(\overline{\boldsymbol{\omega}(\boldsymbol{x}, t) \times \boldsymbol{a}}\right)_{\nu_1 \cdots \nu_k} \equiv$$
$$\equiv D_t^{(k)} \boldsymbol{a} = \left\{\text{all orders}\right\}. \tag{36}$$

Here the *alignment tensor of order l* is defined by

$$a_{\mu_1 \cdots \mu_l}(\boldsymbol{x}, t) := \frac{2l+1}{l!}\int_{S^2} f(\cdot) \ \overline{n_{\mu_1} \cdots n_{\mu_l}} \ d^2n \tag{37}$$

(n_j is the j-th Cartesian component of the microscopic director, and $\overline{n_{\mu_1} \cdots n_{\mu_l}}$ the total symmetric traceless product of them). The right-hand side of (36) has the typical form

$$\left\{ \text{ all orders } \right\} = \mathcal{F}[v(\cdot) - v(x,t), \omega(\cdot) - \omega(x,t)] \tag{38}$$

with the property

$$\mathcal{F}[0,0] = 0 . \tag{39}$$

Therefore (36) can easily be interpreted: Its left hand side is Jaumann's total time derivative for symmetric irreducible tensors. Consequently that side of the equation describes the changes of the alignment tensors which can be observed from a frame translated with $v(x,t)$ and rotated with $\omega(x,t)$. Thus the right hand side of (36) is due to the production of alignment. This production of course will vanish, if there are no deviations in the velocity distributions of $v(\cdot)$ and $\omega(\cdot)$ from their averages $v(x,t)$ and $\omega(x,t)$. The special form of the alignment production involves generally *all* the alignment tensors of any order. A subdynamic even of the alignment tensor of second order will not exist in general. Approximations of (36) are widespread [6].

5. Remark on Constitutive Equations

Because the balance equations include quantities which are determined by the considered material we need constitutive equations for solving the balances. In general there are two different methods to proceed [14]: We can make an ansatz for the constitutive equations, or we can look for a class of materials which is characterized by a chosen state space. In both cases the dissipation inequality (35) has to be exploited. After having inserted the energy balance equation into (34) the expressions for the spin, the stress tensor, and couple force are replaced by their mesoscopic definitions. A straightforward calculation results in a macroscopic dissipation inequality [15] whose shape is different from that of (34). In that dissipation inequality of different shape new macroscopic fields appear which are mesoscopically defined, e.g. the alignment tensor field of second order. After having chosen the state space the transformed dissipation inequality is exploited by the Coleman-Noll procedure [16]. The essential results are [15]:

− the coefficients of the free energy are restricted in comparison to Ericksen-Leslie theory of liquid crystals,
− the entropy flux density is different from heat flux density over temperature,
− the couple stress tensor can be calculated explicitly,
− the Landau condition for equilibrium in liquid crystals can be derived and extended to inhomogeneous alignment.

References

1. Ericksen, J.L. (1960) *Arch. Rat. Mech. Anal.* 4, 231–237.
2. Leslie, J.L. (1968) *Arch. Rat. Mech. Anal.* 28, 265–283.
3. Eringen, A.C. and Lee, J.D. (1974) in J.F. Johnson and R.S. Porter (eds.) *Liquid Crystals and Ordered Fluids, Vol. 2*, Plenum Press, New York, pp. 315–330.
4. Eringen, A.C. (1978) In J.F. Johnson and R.S. Porter (eds.) *Liquid Crystals and Ordered Fluids, Vol. 3*, Plenum Press, New York, pp. 443–474.
5. Hess, S. (1975) *Z. Naturforsch.* 30a, 728–733.
6. Pardowitz, I. and Hess, S. (1980) *Physica* 100A, 540–562.
7. Blenk, S. (1989) Diplomarbeit, Institut für Theoretische Physik, Technische Universität Berlin.
8. Müller, W.H. and Muschik, W. (1983) *J. Non-Equilib. Thermodyn.* 8, 29–46.
9. Muschik, W. and Müller, W.H. (1983) *J. Non-Equilib. Thermodyn.* 8, 47–66.
10. Blenk, S., Ehrentraut, H., and Muschik, W. (1991) *Physica A* 174, 119–138.
11. Blenk, S. and Muschik, W. (1991) *J. Non-Equilib. Thermodyn.* 16, 67–87.
12. Blenk, S., Ehrentraut, H., and Muschik, W. (1991) *Mol. Cryst. Liqu. Cryst.* 204, 133–141.
13. Ehrentraut, H. (1989) Diplomarbeit, Institut für Theoretische Physik, Technische Universität Berlin.
14. Muschik, W. (1990) *Aspects of Non-Equilibrium Thermodynamics*, World Scientific, Singapore, sect. 6.4.
15. Blenk, S., Ehrentraut, H., and Muschik, W. (1992) *Int. J. Engng. Sci.* 30, 1127–1143.
16. Colemann, B.D. and Noll, W. (1963) *Arch. Rat. Mech. Anal.* 13, 167–168.

FINITE-TIME THERMODYNAMICS AND SIMULATED ANNEALING

BJARNE ANDRESEN
Ørsted Laboratory, Niels Bohr Institute
University of Copenhagen
Universitetsparken 5, DK-2100 Copenhagen Ø
Denmark

Abstract.

Finite-time thermodynamics is the extension of traditional reversible thermodynamics to include the extra requirement that the process in question goes to completion in a specified finite length of time. As such it is by definition a branch of irreversible thermodynamics, but unlike most other versions of irreversible thermodynamics, finite-time thermodynamics does not require or assume any knowledge about the microscopics of the processes, since the irreversibilities are described by macroscopic constants such as friction coefficients, heat conductances, reaction rates and the like. Some concepts of reversible thermodynamics, such as potentials and availability, generalize nicely to finite time, others are completely new, e.g. endoreversibility and thermodynamic length. The basic ideas of finite-time thermodynamics are reviewed and several of its procedures presented, emphasizing the importance of power.

The global optimization algorithm simulated annealing was designed for extremely large and complicated systems and therefore inspired by analogy to statistical mechanics. Its basic version is outlined, and several notions from finite-time thermodynamics are introduced to improve its performance. Among these are an optimal temperature path, the use of ensembles, and an analytic two-state model with Arrhenius kinetics.

1. Introduction

1.1 MOTIVATION

From its infancy over 150 years ago thermodynamics has provided limits on work or heat exchanged during real processes. The first problem treated in a systematic way was how much work a steam engine can produce from the burning of one ton of coal. With true scientific generalization Sadi Carnot concluded that any engine taking in heat from a hot reservoir at temperature T_H has to deposit some of that heat in a cold reservoir (e.g. the surroundings), whose temperature we call T_L. The largest fraction of the heat which can be converted into work is

$$\eta_C = 1 - \frac{T_L}{T_H}, \tag{1.1}$$

traditionally known as the Carnot efficiency. This expression contains the two basic ingredients of a thermodynamic limit: (i) it applies to *any* process converting heat into work; and (ii) it is an *absolute* limit, i.e. no process, however ingenious, can do better.

111

J. S. Shiner (ed.), Entropy and Entropy Generation, 111–127.
© 1996 *Kluwer Academic Publishers.*

As thermodynamic theory developed, emphasis changed from process variables like work and heat exchanged to state variables like entropy and chemical potential. A bridge between the two are the thermodynamic work potentials, such as Helmholtz free energy F for isothermal, isochoric processes or the Gibbs free energy G for isothermal, isobaric processes. They are defined such that, under the given conditions, their changes provide upper bounds on the work a process can supply or lower bounds on the work required to drive a process. Gibbs introduced the concept of 'available work' as the maximum work that can be extracted from a system allowed to go from a constrained, internally equilibrated state to a state in equilibrium with its surroundings. This quantity is used more and more frequently in engineering contexts [1, 2] under the names 'availability' in the U.S. and 'exergy' in Europe. For a system relaxing to an ambient temperature T_0, pressure P_0, and chemical potentials μ_{0i} it is given by

$$A = U + P_0V - T_0S - \sum_i \mu_{0i}N_i \qquad (1.2)$$

and is thus not a state function in the usual sense of depending only on variables of the system; the availability depends on the intensive variables of the environment as well.

Such criteria of merit have long been common currency for thermodynamic studies in physics, chemistry, and engineering. They all share one characteristic: The ideal to which any real process is compared is a *reversible process*. Stated in a different way, *traditional thermodynamics is a theory about equilibrium states and about limits on process variables for transformations from one equilibrium state to another.* Nowhere does time enter the formulation, so these limits must be the lossless, reversible processes which proceed infinitely slowly and thus take infinite length of time to complete. However, referring back to the original question addressed by Carnot, who is interested in an engine which operates infinitely slowly (and thus produces zero power) — or any other process with zero rate of operation, for that matter?

In order to obtain more realistic limits to the performance of real processes *finite-time thermodynamics is designed as the extension of traditional thermodynamics to deal with processes which have explicit time or rate dependencies.* These constraints, of course, imply a certain amount of loss, or entropy production, which is at the heart of the question posed above.

1.2 EARLY DEVELOPMENTS

In the course of developing finite-time thermodynamics we discovered that a few isolated papers already had considered different aspects of processes operating at nonzero rates. The first of these was the important work of Tolman and Fine [3] who put the Second Law of thermodynamics into equality form,

$$W = \Delta A - T_0 \int_{t_i}^{t_f} \dot{S}_{tot}\, dt \qquad (1.3)$$

by subtracting the work equivalent of the entropy produced during the process from the reversible work, i.e. the decrease of system availability, as defined in eq. (1.2). The superscript dot indicates rate, and the integral limits are the initial and final times of the process. This is a quantification of the 'price of haste'.

Another model has evolved into almost a classic paradigm of systems operating in finite time. This is the model of Curzon and Ahlborn [4], a Carnot engine with the simple constraint that it be linked to its surroundings through *finite* heat conductances. Figure 1 illustrates the slightly more general endoreversible system with the triangle signifying any reversible engine. (The term endoreversible means 'internally reversible', i.e. all irreversibilities reside in the coupling of flows to the surroundings. In this case the irreversibilities are the resistance to heat transfer and possibly friction.) It turns out that the results derived by Curzon and Ahlborn explicitly for an interior Carnot engine are equally valid for a general endoreversible system. The maximum efficiency of their engine is of course $\eta_C = 1 - T_L/T_H$, obtained at zero rate so that losses across the resistors vanish, but these authors showed that, when the system operates to produce *maximum power*, the efficiency of the engine is only

$$\eta_w = 1 - \sqrt{\frac{T_L}{T_H}}. \qquad (1.4)$$

Besides the simplicity of the expression it is remarkable that it does not contain the value of the heat conductances.

2. Performance Bound Without Path

The smallest amount of information one can ask for concerning the performance of a system is a single number, e.g. the work or heat exchanged during the process, its efficiency, or any other figure of merit. In most cases this can be calculated without knowledge of the detailed path followed and is then computationally much simpler to obtain.

2.1 GENERALIZED POTENTIALS

In traditional thermodynamics potentials are used to describe the ability of a system to perform some kind of work under given constraints. These constraints are usually the constancy of some state variables like pressure, volume, temperature, entropy, chemical potential, particle number, etc. Under such conditions the decrease in thermodynamic potential P from state i to state f is equal to the amount of work that is produced when a reversible process carries out the transition, and hence is the upper bound to the amount of work produced by any other process,

$$W \le W_{rev} = P_i - P_f. \qquad (2.1)$$

In this section we will show that the constraints need not simply be the constancy of some state variable, and that the potentials may be generalized to contain constraints involving time [5]. The procedure will be a straight forward extension of the Legendre transformations [6] used in traditional thermodynamics [7, 8], and we will start with such an example.

In a reversible process heat and work can be expressed as inexact differentials,

$$dQ = TdS, \qquad dW = PdV, \qquad (2.2)$$

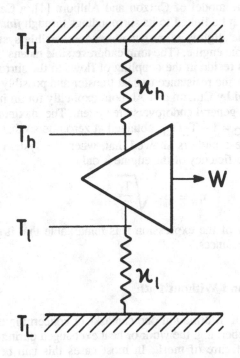

Figure 1. An endoreversible engine has all its losses associated with its coupling to the environment, there are no internal irreversibilities. This is illustrated here as resistances in the flows of heat to and from the working device indicated by a triangle. These unavoidable resistances cause the engine proper to work across a smaller temperature interval, $[T_h;T_l]$ than that between the reservoirs, $[T_H;T_L]$, one which depends on the rate of operation.

i.e. they cannot by themselves be integrated, further constraints defining the integration path are required. Such a constraint could be that the process is isobaric, $dP = 0$. One can then add a suitable integrating zero-term, xdP to make dW an exact differential. The obvious choice is $x = V$,

$$dW = PdV = PdV + VdP = d(PV), \tag{2.3}$$

such that the isobaric work potential becomes $P = PV$. Similarly the isobaric heat potential $U + PV$ is obtained from

$$dQ = TdS = dU + PdV = dU + PdV + VdP = d(U + PV), \tag{2.4}$$

where the combined First and Second Laws of thermodynamics

$$dU = TdS - PdV \tag{2.5}$$

have been used.

Now, the constraints need not be the constancy of one of the state variables. Consider a balloon with constant surface tension α. In equilibrium with an external pressure P_{ex} such a sphere of radius r has an internal pressure

$$P = P_{ex} + \frac{2\alpha}{r}, \qquad (2.6)$$

which can be rearranged into

$$(P - P_{ex})V^{1/3} = 2\alpha \left(\frac{4\pi}{3}\right)^{1/3}. \qquad (2.7)$$

Since the right hand side of this equation is a constant, this means that $(P - P_{ex})V^{1/3}$ is an integral of motion for the fluid inside the balloon. We can then add a suitable amount of $d[(P - P_{ex})V^{1/3}]$ $(=0)$ to dW to make it exact [see eq. (2.11) below for how to do it],

$$dW = PdV = PdV + \frac{3}{2}V^{2/3} d[(P - P_{ex})V^{1/3}] \qquad (2.8)$$

$$= d[\tfrac{1}{2}V(3V - P_{ex})].$$

Thus the work done by the coupled system, surface + fluid, is given by the decrease in the potential $P = \frac{1}{2}V(3V - P_{ex})$, regardless of path followed.

In its most general form the Legendre transformation can be used to calculate a potential P for the arbitrary process variable B, expressible as a path integral in terms of generalized forces f_i and displacements x_i,

$$B = \sum_i \int f_i\, dx_i = \int f \cdot dx, \qquad (2.9)$$

B will usually be work, and vector notation is used for compactness. To find P, one adds to $f \cdot dx$ an integrating term $g \cdot dy$, where dy is necessarily zero as a result of the constraints defining the process. Note that $dy = 0$ may involve time and could come from a condition in the form of a differential equation as well as from the more familiar thermodynamic condition of a constant variable, as used in the example above. Hence the differential form $dy = 0$ is used rather than the integrated form $y = \mathbf{constant}$, since y itself may not exist. The mathematical problem of finding P has two steps, finding a function g which makes $d\omega = f \cdot dx + g \cdot dy$ an exact differential dP, and then integrating to get P itself. The first step involves the Cauchy-Riemann condition that $d\omega$ has equal cross derivatives with respect to the free state variables, e.g. a and b:

$$\frac{\partial}{\partial b}\left[f \cdot \left(\frac{\partial x}{\partial a}\right)_b + g \cdot \left(\frac{\partial y}{\partial a}\right)_b \right] = \frac{\partial}{\partial a}\left[f \cdot \left(\frac{\partial x}{\partial b}\right)_a + g \cdot \left(\frac{\partial y}{\partial b}\right)_a \right] \qquad (2.10)$$

or

$$\left(\frac{\partial g}{\partial a}\right)_b \cdot \left(\frac{\partial y}{\partial b}\right)_a - \left(\frac{\partial g}{\partial b}\right)_a \cdot \left(\frac{\partial y}{\partial a}\right)_b = \left(\frac{\partial f}{\partial b}\right)_a \cdot \left(\frac{\partial x}{\partial a}\right)_b - \left(\frac{\partial f}{\partial a}\right)_b \cdot \left(\frac{\partial x}{\partial b}\right)_a. \qquad (2.11)$$

With f, dx, and dy known, this is the equation from which g may be obtained. In the usual case of $f = P$, $dx = dV$, $a = V$, and $b = P$, the right hand side of eq. (2.11) simplifies to 1. The second step in finding P, the integration of dP, is, of course, only unique within a constant of the motion; i.e. two methods of integration may yield two different potentials P and P', but their variations will always be the same, $\Delta P = \Delta P'$.

2.2 THERMODYNAMIC LENGTH

In an effort to develop a more direct and transparent way of calculating all the usual partial derivatives in traditional thermodynamics Weinhold [9–13] proposed using scalar products between vectors in the abstract space of equilibrium states of a system, represented by all its extensive variables X_i. The products were defined relative to

$$M_U = \left\{ \frac{\partial^2 U}{\partial X_i \partial X_j} \right\} \qquad (2.12)$$

as the metric, where U is the internal energy. However, second derivatives are usually identified as curvatures and, as such, should be interpreted as availabilities in Gibbs space (U as a function of all the other extensive variables). This lead us to seek another interpretation of pathlengths calculated with Weinhold's metric, now called thermodynamic lengths, and we [14] found that they are always changes in some molecular velocities, depending, of course, on the constraints of the process (isobaric, isochoric, etc.).

Subsequently Salamon and Berry [15] found a connection between the thermodynamic length along a process path and the (reversible) availability lost in the process. Specifically, if the system moves via states of local thermodynamic equilibrium from an initial equilibrium state i to a final equilibrium state f in time τ, then the dissipated availability $-\Delta A$ is bounded from below by the square of the distance (i.e. length of the shortest path) from i to f times ε/τ, where ε is a mean relaxation time of the system. If the process is endoreversible, the bound can be strengthened to

$$-\Delta A \geq \frac{L^2 \varepsilon}{\tau}, \qquad (2.13)$$

where L is the length of the *traversed* path from i to f. Equality is achieved at constant thermodynamic speed $v = dL/dt$ corresponding to a temperature evolution given by [16, 17]

$$\frac{dT}{dt} = -\frac{vT}{\varepsilon\sqrt{C}}, \qquad (2.14)$$

where C is the heat capacity of the system. For comparison, the bound from traditional thermodynamics is only

$$-\Delta A \geq 0. \qquad (2.15)$$

An analogous expression exists for the total entropy production during the process:

$$\Delta S^u \geq \frac{L^2 \varepsilon}{\tau}. \tag{2.16}$$

The length L is then calculated relative to the entropy metric

$$M_S = -\left\{ \frac{\partial^2 S}{\partial X_i \partial X_j} \right\} \tag{2.17}$$

which (when expressed in identical coordinates!) is related to M_U by [18]

$$M_U = -T_0 M_S, \tag{2.18}$$

where T_0 as usual is the environment temperature. In statistical mechanics, where entropy takes the form

$$S(\{p_i\}) = -\sum_i p_i \ln p_i, \tag{2.19}$$

the metric M_S is particularly simple, being the diagonal matrix [19]

$$M_S = -\left\{ \frac{1}{p_i} \right\} \tag{2.20}$$

The same procedure of calculating metric bounds for dynamic systems has been applied to coding of messages [20] and to economics [21].

More recently [22] we have relaxed a number of the assumptions in the original work, primarily those restricting the system to be close to equilibrium at all times and the average form of the relaxation time ε. The more general bound replacing eq. (2.16) then becomes

$$\Delta S^u \geq \frac{1}{\Xi} \left(\int_{\xi_i}^{\xi_f} \frac{1}{T\sqrt{C}} \left| \frac{dU}{d\xi} \right| \sqrt{1 + \frac{\theta}{CT} \frac{dU}{d\xi} + \ldots} \; d\xi \right)^2 \tag{2.21}$$

with $\Xi = \xi_f - \xi_i$ being the total duration of the process in natural dimensionless time units,

$$d\xi = dt/\varepsilon(T), \tag{2.22}$$

and where we have defined

$$\theta(T) = 1 + \frac{T}{2} \frac{\partial C}{C \; \partial T}. \tag{2.23}$$

The equality (lower bound) in eq. (2.21) is achieved when the integrand is a constant, i.e. when

$$\frac{dS^u}{d\xi} = \text{constant} . \tag{2.24}$$

Consequently, constant rate of entropy production, when expressed in terms of natural time, is the path or operating strategy which produces the least overall entropy.

One can express the optimal path, eq. (2.24), in a form similar to eq. (2.14):

$$\frac{dT}{dt} \sqrt{1 + \frac{\theta(T) \, \varepsilon \, (dT/dt)}{T} + \dots} = \text{constant} \times \frac{T}{\varepsilon \sqrt{C}}. \tag{2.25}$$

The constant thermodynamic speed algorithm, eq. (2.14), is thus the leading term of the general solution in an expansion about equilibrium behavior.

3. Optimal path

3.1 OPTIMAL PATH CALCULATIONS

A knowledge of the maximum work that can be extracted during a given process, e.g. calculated by one of the procedures described in the previous section, may not by itself be sufficient. One may also want to know *how* this maximum work can be achieved, i.e. the time path of the thermodynamic variables of the system. The primary tool for obtaining this path is optimal control theory.

This is not the place to repeat the mechanics of optimal control calculations (see e.g. [23–25]). Let it suffice to point out that in order to set up the optimal control problem one must specify

- the controls, i.e. the variables that can be manipulated by the operator (they may be a volume, rate, voltage, heat conductance, etc.);
- limits on the controls and on the state variables, if any (in order to avoid unphysical situations such as negative temperatures and infinite speeds);
- the equations that govern the time evolution of the system (they will usually be differential equations describing heat transfer rates, chemical reaction rates, friction, and other loss mechanisms);
- the constraints that are imposed on the system (e.g. conserved quantities, the quantities held constant, or any requirements on reversibility. The constraints may either be differential, instantaneous i.e. algebraic, or integral i.e. not obeyed at each point but over the entire interval);
- the desired quantity to be maximized, called the objective function (usually expressed as an integral); and finally
- whether the duration of the process is fixed or part of the optimization.

Typical manipulation usually leads to a set of coupled, non-linear differential equations for which a qualitative analysis and a numerical solution are the only hope. Thus answering the more demanding question about the optimal time path rather than the standard question about maximum performance requires a considerably larger computational effort. On the other hand, once the time path is calculated, all other thermodynamic quantities may be calculated from it, much like the wave function is the basis of all information in quantum mechanics.

3.2 CRITERIA OF PERFORMANCE

Efficiency, the earliest criterion of performance for engines, measured how much water could be pumped out of a mine by burning a ton of coal. Other familiar criteria include

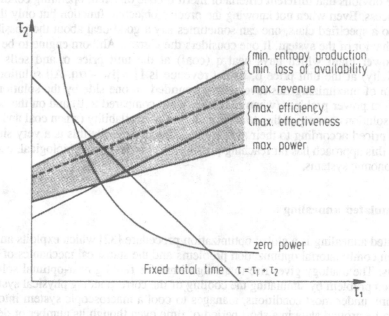

Figure 2. If an endoreversible engine (Fig. 1) spends time τ_1 in contact with the hot reservoir and τ_2 in contact with the cold reservoir, the optimal proportioning between τ_1 and τ_2 depends on what one chooses to optimize, as indicated. The locus of maximum revenue for a power producing system falls in the shaded area for any choice of prices, as described in the text. Only contact times above the hyperbola marked 'zero power' actually correspond to positive power production.

effectiveness (efficiency relative to the Carnot efficiency), change of thermodynamic potential, and loss of availability, all of which are measures of work. Potentials for heat can also be defined (see Sect. 2.1) but are less common, and we have used the minimization of entropy production in a separate study [26].

The Curzon-Ahlborn analysis and most of our own analyses use a quite different criterion, that of *power*. This quantity is of course zero for any reversible system, and maximizing power forces us to deal with systems operating at finite rates. Other criteria of performance are the rate of entropy production and the rate of loss of availability. Entropy production was a function introduced in the earliest thinking about irreversible thermodynamics [27–30], but more from the differential, local, instantaneous viewpoint than from the global, integral view of entire optimized processes. Under some circumstances, optimizing one of these quantities is equivalent to optimizing another. For example, minimizing the entropy production is equivalent to minimizing the loss of availability, at least in those cases in which the irreversibilities can be represented as spontaneous heat flows [26].

Salamon and Nitzan [31] have optimized the Curzon-Ahlborn engine for a number of these objective functions. Assuming the working fluid to be in contact with the hot reservoir for the period τ_1 and the cold reservoir for the period τ_2, the optimal time distributions are shown in Fig. 2. The diagonal $\tau = \tau_1 + \tau_2$ indicates fixed total cycle time, and only processes above the curve labeled 'zero power' produce positive power. It is quite obvious that different criteria of merit dictate different operating conditions for the process. Even when not knowing the precise objective function but only that it belongs to a specified class, one can sometimes say a good deal about the possible optimal behavior of the system. If one considers the Curzon-Ahlborn engine to be a model of a power plant which buys heat q (coal) at the unit price α and sells work w (electricity) at the unit price β, its net revenue is $\Pi = \beta w - \alpha q$. All solutions to the problem of maximizing this revenue are bounded on one side by the solutions to the maximum power problem (when α is significant compared to β) and on the other side by the solutions corresponding to minimum loss of availability (when coal and electricity are priced according to their availability contents). While this is a very simple example, this approach has far reaching possibilities for describing biological, ecological, and economic systems.

4. Simulated annealing

Simulated annealing is a global optimization procedure [32] which exploits an analogy between combinatorial optimization problems and the statistical mechanics of physical systems. The analogy gives rise to an algorithm for finding near-optimal solutions to the given problem by simulating the cooling of the corresponding physical system. Just as nature, under most conditions, manages to cool a macroscopic system into or very close to its ground state in a short period of time even though its number of degrees of freedom is of the order of Avogadro's number, so does simulated annealing rapidly find a good guess of the solution of the posed problem.

Even though the original class of problems under consideration [32] was combinatorial optimization, excellent results have been obtained with seismic inversion [33], pattern recognition [34], and neural networks [35] as well, just to name a few. Since layout and scheduling problems in computer and telecommunications systems frequently contain elements of combinatorial optimization, and their complexity often rivals and even surpasses that of a mole of chemical substance, simulated annealing is excellently suited to attack those problems. The virtue of simulated annealing is its efficiency as a general-purpose method for handling extremely complicated problems for which no direct solution method is known without the need for developing an ad hoc algorithm.

4.1 THE ALGORITHM

The simulated annealing algorithm is based on the Monte Carlo simulation of physical systems. It requires the definition of a state space $\Omega = \{\omega\}$ with an associated cost function (physical analog: energy) $E: \Omega \rightarrow R$ which is to be minimized in the optimization. At each point of the Monte Carlo random walk in the state space the system may make a jump to a neighboring state; this set of neighbors, known as the move class $N(\omega)$, must of course also be specified. Alternatively one may specify the complete transition probability matrix $\mathbf{P}$ between all states of the system. The only control parameter of the algorithm is the temperature T of the heat bath in which the corresponding physical system is immersed, measured in energy units, i.e. we take Boltzmann's constant $k = 1$.

The random walk inherent in the Monte Carlo simulation is accomplished by the Metropolis algorithm [36] which states that:

(i) At each step t of the algorithm a neighbor ω' of the current state ω_t is selected at random from the move class $N(\omega_t)$ to become the candidate for the next state.

(ii) It actually becomes the next state only with probability

$$P_{accept} = \begin{cases} 1 & \text{if } \Delta E \leq 0 \\ e^{-\Delta E/T_t} & \text{if } \Delta E > 0 \end{cases}, \tag{4.1}$$

where $\Delta E = E(\omega') - E(\omega_t)$ is the increase in cost for the move. If this candidate is accepted, then $\omega_{t+1} = \omega_t$.

The only thing left to specify is the sequence of temperatures T_t appearing in the Boltzmann factor in P_{accept}, the so-called annealing schedule. Like in metallurgy, this cooling rate has a major influence on the final result. A quench is quick and dirty, often leaving the system stranded in metastable states high above the ground state/optimal solution. Slow annealing produces the best result but is computationally expensive.

This completes the formal definition of the simulated annealing algorithm, which in principle simply is repeated numerous times until a satisfactory result is obtained.

4.2 THE OPTIMAL ANNEALING SCHEDULE

So far all suggested simulated annealing temperature paths (annealing schedules) have been of the a priori type and thus have not adjusted to the actual behavior of the system as the annealing progresses. Examples of such schedules are

$$T(t) = a\, e^{-t/b} \tag{4.2}$$

$$T(t) = \frac{a}{b+t} \tag{4.3}$$

$$T(t) = \frac{a}{\ln(b+t)}. \tag{4.4}$$

The real annealing of physical systems often has rough parts where the surrounding temperature must be decreased slowly due to phase transitions or regions of large heat capacity or slow internal relaxation. The same behavior is seen in the abstract systems, so annealing schedules which take such variations into account are preferable in order to keep computation time at a minimum for a given accuracy of the final result [37].

At this point I would like to make the analogy between the abstract simulated annealing process and a real thermodynamic process even stronger. Specifically, if the correspondence with statistical mechanics implied in the simulated annealing procedure involving phase space and state energies is valid, then further results from thermodynamics will probably also carry over to simulated annealing.

Since asking a question (= one evaluation of the energy function) in information theoretic terms is equivalent to producing one bit of entropy, the computationally most efficient procedure will be the temperature schedule $T(t)$ which overall produces minimum entropy. Above we have derived various bounds and optimal paths for real thermodynamic systems using finite-time thermodynamics [15, 19, 26, 38]. The minimum-entropy production path which most readily generalizes to become the optimal simulated annealing schedule, is the one calculated with thermodynamic length [cf. eq. (6.4) below]:

$$\frac{dT}{dt} = -\frac{vT}{\varepsilon\sqrt{C}} \tag{2.14}$$

or equivalently

$$\frac{\langle E \rangle - E_{eq}(T)}{\sigma} = v. \tag{4.5}$$

As long as the system in question is accurately described by statistical mechanics, these bounds and the paths that achieve them will provide the most efficient solution within the allotted time. Since the layout of major networks and the allocation of access to common scarce resources are closely related to graph partitioning, simulated annealing with a constant thermodynamic speed schedule should yield good results. In these expressions v is the (constant) thermodynamic speed, C and ε are the heat capacity and internal relaxation time of the system, respectively, $\langle E \rangle$ and σ the corresponding mean energy and standard deviation of its natural fluctuations, and finally $E_{eq}(T)$ is the internal energy the system would have if it were in equilibrium with its surroundings at temperature T. The physical interpretation of eq. (4.5) is that the environment should at all times be kept v standard deviations ahead of the system. Similarly eq. (2.14) indicates that the annealing should slow down where internal relaxation is slow and where large amounts of 'heat' has to be transferred out of the system [17]. In case C and ε do not vary with temperature, eq. (2.14) integrates to the standard schedule eq. (4.2). The more realistic assumption of an Arrhenius-type relaxation time, $\varepsilon \sim \exp(a/T)$, and a heat capacity $C \sim T^{-2}$ implies the vastly slower schedule eq. (4.4). Reality is usually in between these extremes. Figure 3 shows the successive decrease in energy for annealings on a graph partitioning problem following different annealing schedules.

4.3 PARALLEL IMPLEMENTATION

The extra temperature dependent variables of the constant thermodynamic speed schedule of course require additional computational effort. Since systems often change considerably in a few steps, ergodicity is not fulfilled, so the use of time averages to obtain $\langle E \rangle$, σ, C, and ε is usually not satisfactory. Instead we suggest [39] running an ensemble of systems in parallel, i.e. with the same annealing schedule, in the true spirit of the analogy to statistical mechanics. Then these variables can be obtained anytime as ensemble averages based on the system degeneracies $p_i = p(E_i)$:

$$Z(T) = \sum_i p_i \exp(-E_i/T) \tag{4.6}$$

$$E(T) = T^2 \frac{d \ln Z}{dt} \tag{4.7}$$

$$C(T) = \frac{dE}{dT} = \frac{\langle (\Delta E)^2 \rangle}{T^2} \tag{4.8}$$

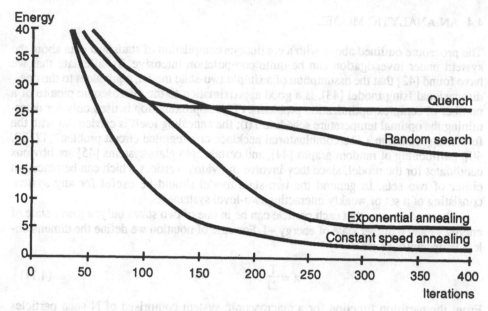

Figure 3. Energy for a graph partitioning problem as the annealing progresses, using different annealing schedules: quench (T=0), random search (T=∞), exponential (eq. (4.2)), constant speed (eq. (2.14) or (4.5)).

$$\varepsilon(T) = \frac{-1}{\ln\lambda_2} \approx \frac{T^2 \, C(T)}{\sum_i p_i \sum_{j>i} (E_j - E_i)^2 \, P_{ji} \, \exp(-E_i/T)}, \qquad (4.9)$$

where λ_2 is the second largest eigenvalue of the thermalized version of the transition probability matrix **P** among all the energy levels ($\lambda_1 = 1$ corresponds to equilibrium).

But from where does one get the degeneracies p_i? Actually [39], information to calculate the temperature-independent (or infinite-temperature, if you prefer) transition probability matrix **P** can be accumulated during the annealing run by simply adding up in a matrix **Q** the number of attempted moves (not just the accepted ones) from level i to j as the calculation progresses. Normalization of **Q** yields a good estimate of **P**,

$$P_{ji} = Q_{ji} \Big/ \sum_k Q_{ki}. \qquad (4.10)$$

The degeneracies **p** are then the eigenvector of **P** corresponding to the eigenvalue 1.

This use of ensemble annealing is particularly well suited for implementation on present day parallel computers. A further analysis of its performance has been carried out by Ruppeiner, Pedersen, and Salamon [40], and the trade-off between ensemble size and duration of annealing for fixed total computation cost has been addressed by Pedersen et al. [41].

4.4 AN ANALYTIC MODEL

The procedure outlined above with a continuous compilation of statistical data about the system under investigation can be quite computation intensive. To alleviate that we have found [42] that the assumption of a simple two-state model, equivalent to the one-dimensional Ising model [43], is a good approximation to the microscopic picture of a number of complex optimization problems. (This approximation is used only for determining the optimal temperature schedule T(t); the annealing itself is carried out with the full energy function.) The combinatorial necklace or integrated circuit problem [17, 39, 40], partitioning of random graphs [44], and certain spin glass systems [45] are obvious candidates for the model, since they involve individual vertices which can be placed in either of two sets. In general the two-state model should be useful for any system consisting of a set of weakly interacting two-level systems.

The model states that each particle can be in one of two states only: a lower state of energy −J and a higher state of energy +J. For ease of notation we define the dimensionless temperature variable

$$x = \frac{T}{2J} .$$

(4.11)

From the partition function for a macroscopic system comprised of N such particles (N >> 1) at temperature x the system energy and heat capacity follow directly:

$$E(x) = - NJ \tanh \frac{1}{2x}$$

(4.12)

$$C(x) = \frac{N \left(\frac{1}{2x}\right)^2}{\cosh^2 \frac{1}{2x}} .$$

(4.13)

In addition we assume that the system relaxation time is well represented by the classical Arrhenius expression

$$\varepsilon(x) = A \exp(B/x),$$

(4.14)

where A is a (constant) collision frequency, and B is the apparent barrier height of the transition state. Even though the energy landscape of the system will generally contain several different barriers, only the highest will be effective in the long-time limit when all the faster relaxations have died out, i.e. close to equilibrium. This assumption is consistent with the derivation of eq. (2.14).

Introducing these assumptions into the optimal rate annealing schedule eq. (2.14) yields

$$\frac{dx}{dt} = -v' \, x^2 \left(\exp\left(\frac{1}{x}\right) + 1 \right) \exp\left(-\frac{B+1/2}{x}\right) ,$$

(4.15)

where v' is a constant. This is the differential equation defining the optimal temperature path x(t) for a two-state model with an Arrhenius-type relaxation. It can, at least in principle, be solved analytically before any annealing begins and thus replaces the much more time consuming procedure of collecting statistical information along the way presented in the previous section [39]. The 'price' for this faster procedure clearly is less generality and required knowledge about the system in advance.

5. Summary

Finite-time thermodynamics was 'invented' in 1975 by R. S. Berry, P. Salamon, and myself as a consequence of the first world oil crisis. It simply dawned on us that all the existing criteria of merit were based on reversible processes and therefore were totally unrealistic for most real processes. That made an evaluation of the potential for improvement of a given process quite difficult.

Finite-time thermodynamics is developed from a macroscopic point of view with heat conductances, friction coefficients, overall reaction rates, etc. rather than based on a microscopic knowledge of the processes involved. Consequently most of the ideas of traditional thermodynamics have been assimilated, e.g. the notions of thermodynamic potential (Sect. 2.1) and availability. At the same time we have seen new concepts emerge, e.g. the non-equivalence of well-honored criteria of merit (Sect. 3.2), the importance of power as the objective (Sect. 1.2 and 3.2), the generality of the endoreversible engine, and in particular thermodynamic length (Sect. 2.2). Several of these abstract concepts have been successfully applied to practical optimizations [46–52].

Lately the notions and results of finite-time thermodynamics, at times in connection with statistical mechanics and information theory, have been used to perfect the global optimization method simulated annealing (Sect. 4). However, the surface has only been scratched, there is still plenty of room for inspiration from such well-known concepts as state entropy, free energy, and transition state theory. Simulated annealing has proven to be a very useful general purpose optimization algorithm for extremely complicated problems, even with fixed temperature schedules, eqs. (4.2) – (4.4). Elaborating the analogy to physical statistical mechanical systems with the introduction of optimality results from finite-time thermodynamics has improved the efficiency of the algorithm noticeably, and the use of ensembles of random walkers has made it self-adapting. Finally, a simple analytical model based on the one-dimensional Ising model may reduce the computational time considerably.

The method as described above mimics nature as it equilibrates systems on an atomic or molecular scale (e.g. chemical reactions or crystal structures). Inspiration can also be lifted from nature's way of developing the most favorable systems on the macroscopic biological scale: evolution. Several annealing procedures based on evolution with its cloning of successful species and dying out of unsuccessful ones have been developed as replacements for the Metropolis algorithm of Sect. 4.1, notably by Ebeling and coworkers [53–55] and by Schuster [56]. In my opinion the two lines of thought are complementary in the sense that they are advantageous for many trials and for a more limited set of tests, respectively.

Acknowledgments

Much of the work reported in this review has been done in collaboration with colleagues near and far. In particular I want to express my gratitude to Profs. R. Stephen Berry, Peter Salamon, Jim Nulton, Karl Heinz Hoffmann, Jeff Gordon, and Jacob Mørch

Pedersen. Preparation of this manuscript has been supported by the European Union under the Human Capital and Mobility program (contract number ER-BCHRXCT920007).

References

1. Keenan, J.H. (1941) *Thermodynamics*, MIT Press, Cambridge.
2. Gaggioli, R.A. (ed.) (1980) *Thermodynamics: Second Law Analysis*, *ACS Symposium Series* 122, American Chemical Society, Washington.
3. Tolman, R.C. and Fine, P.C. (1948) *Rev. Mod. Phys.* 20, 51-77.
4. Curzon, F.L. and Ahlborn, B. (1975) *Am. J. Phys.* 43, 22-24.
5. Salamon, P., Andresen, B. and Berry, R.S. (1977) *Phys. Rev. A* 15, 2094-2102.
6. Hermann, R. (1973) *Geometry, Physics, and Systems*, Marcel Dekker, New York.
7. Callen, H.B. (1985) *Thermodynamics*, Wiley, New York.
8. Tisza, L. (1966) *Generalized Thermodynamics*, MIT Press, Cambridge.
9. Weinhold, F. (1975) *J. Chem. Phys.* 63, 2479-2482.
10. Weinhold, F. (1975) *J. Chem. Phys.* 63, 2484-2487.
11. Weinhold, F. (1975) *J. Chem. Phys.* 63, 2488-2495.
12. Weinhold, F. (1975) *J. Chem. Phys.* 63, 2496-2501.
13. Weinhold, F. (1978) in D. Henderson and H. Eyring (eds.) *Theoretical Chemistry, Advances and Perspectives*, Vol. 3, Academic Press, New York, pp. 15-54.
14. Salamon, P., Andresen, B., Gait, P.D. and Berry, R.S. (1980) *J. Chem. Phys.* 73 1001-1002; erratum *ibid.* 73, 5407E.
15. Salamon, P. and Berry, R.S. (1983) *Phys. Rev. Lett.* 51, 1127-1130.
16. Nulton, J. and Salamon, P. (1988) *Phys. Rev. A* 37, 1351-1356.
17. Salamon, P., Nulton, J., Robinson, J., Pedersen, J.M., Ruppeiner, G. and Liao, L. (1988) *Comp. Phys. Comm.* 49, 423-428.
18. Salamon, P., Nulton, J. and Ihrig, E. (1984) *J. Chem. Phys.* 80, 436-437.
19. Feldmann, T., Andresen, B., Qi, A. and Salamon, P. (1985) *J. Chem. Phys.* 83, 5849-5853.
20. Flick, J.D., Salamon, P. and Andresen, B. (1987) *Info. Sci.* 42, 239-253.
21. Salamon, P., Komlos, J., Andresen, B. and Nulton J. (1987) *Math. Soc. Sci.* 13, 153-163.
22. Andresen, B. and Gordon, J. M. (1994) *Phys. Rev. E* 50, 4346-4351.
23. Boltyanskii, V.G. (1971) *Mathematical Methods of Optimal Control*, Holt, Reinhart, and Winston, New York.
24. Tolle, H. (1975) *Optimization Methods*, Springer, New York.
25. Leondes, C.T. (ed.) (1964 and onward). *Advances in Control Systems*, Academic Press, New York.
26. Salamon, P., Nitzan, A., Andresen, B. and Berry, R.S. (1980) *Phys. Rev. A* 21, 2115-2129.
27. Onsager, L. (1931) *Phys. Rev.* 37, 405-426.
28. Onsager, L. (1931) *Phys. Rev.* 38, 2265-2279.
29. Prigogine, I. (1962) *Non-equilibrium Statistical Mechanics*, Wiley, New York.
30. Prigogine, I. (1967). *Introduction to Thermodynamics of Irreversible Processes*, Wiley, New York, pp. 76 ff.
31. Salamon, P. and Nitzan, A. (1981) *J. Chem. Phys.* 74, 3546-3560.
32. Kirkpatrick, S., Gelatt, C.D. Jr. and Vecchi, M.P. (1983) *Science* 220, 671-680.
33. Jakobsen, M.O., Mosegaard, K. and Pedersen, J.M. (1987) in A. Vogel (ed.) *Model Optimization in Exploration Geophysics 2. Proceedings of the 5th International Mathematical Geophysics Seminar*, Vieweg & Sohn, Braunschweig.
34. Hansen, L.K. (1990) At *Symposium on Applied Statistics*, (UNI-C, Copenhagen).
35. Hansen, L.K. and Salamon, P. (1990) *IEEE Trans. on Pattern Analysis and Machine Intelligence* 12, 993-1001.
36. Metropolis, N., Rosenbluth, A.W., Rosenbluth, M.N., Teller, A.H. and Teller, E. (1953) *J. Chem. Phys.* 21, 1087-1092.
37. Ruppeiner, G. (1988) *Nucl. Phys. B (Proc. Suppl.)* 5A, 116-121.
38. Andresen, B. (1983). *Finite-time Thermodynamics*, Physics Laboratory II, University of Copenhagen.
39. Andresen, B., Hoffmann, K.H., Mosegaard, K., Nulton, J., Pedersen, J.M. and Salamon, P. (1988) *J. Phys. (France)* 49, 1485-1492.
40. Ruppeiner, G., Pedersen, J.M. and Salamon, P. (1991) *J. Phys. I* 1, 455-470.
41. Pedersen, J.M., Mosegaard, K., Jacobsen, M.O. and Salamon, P. (1989) Physics Laboratory, University of Copenhagen, Report no. 89-14.
42. Andresen, B. and Gordon, J.M. (1993) *Open Syst. and Info. Dyn.* 2, 1-12.
43. Kubo, R. (1967) *Statistical Mechanics*, North-Holland, Amsterdam.
44. Fu, Y. and Anderson, P.W. (1986) *J. Phys. A* 19, 1605-1620.
45. Grest, G.S., Soukoulis, C.M. and Levin, K. (1986) *Phys. Rev. Lett.* 56, 1148-1151.
46. Hoffmann, K.H., Watowitch, S. and Berry, R.S. (1985) *J. Appl. Phys* 58, 2125-2134.
47. Mozurkewich, M. and Berry, R.S. (1981) *Proc. Natl. Acad. Sci. USA* 78, 1986.

48. Mozurkewich, M. and Berry, R.S. (1982) *J. Appl. Phys* **53**, 34-42.
49. Gordon, J.M. and Zarmi, Y. (1989) *Am. J. Phys.* **57**, 995-998.
50. Andresen, B., Gordon, J.M. (1992) *J. Appl. Phys.* **71**, 76-79.
51. Andresen, B. and Gordon, J.M. (1992) *Int. J. Heat and Fluid Flow* **13**, 294-299.
52. Gordon, J.M. and Ng, K.C. (1994) *J. Appl. Phys.* **57**, 2769-2774.
53. Boseniuk, T. and Ebeling, W. (1988) *Europhys. Lett.* **6**, 107.
54. Boseniuk, T. and Ebeling, W. (1988) *Syst. Anal. Model. Simul.* **5**, 413.
55. Ebeling, W. (1990) *Syst. Anal. Model. Simul.*. **7**, 3.
56. Fontana, W., Schnabl, W. and Schuster, P. (1989) *Phys. Rev. A* **40**, 3301-3321.

ENTROPY, INFORMATION AND
BIOLOGICAL MATERIALS

G.H.A. COLE

University of Hull, Hull, HU6 7RX, England

Abstract.

The requirements of the application of thermodynamics to the Universe as a whole are met by an hypothesis which makes animate materials as universal as inanimate materials throughout the Universe. This places biological materials in a universal evolutionary sequence.

1. Introduction

Living material is virulent and tenacious on Earth and has adapted to the full range of physical conditions. It is an ordered complex of components which, for a limited time (the life time of an individual example), is not in thermal equilibrium with its surroundings [1]. It is a self-reproducing engine that accepts and rejects free energy either from solar radiation or from the internal heat of the Earth (for example the limited life forms surrounding mid-ocean ridges): the absorption may be direct (photosynthesis) or indirect (mineral acquisitions through "food") [2]. The central molecular structure and dynamics in all cases is that of DNA, variations arising from variants of the DNA structure. The relative distributions of those chemical elements making up animate matter are closely the same as for inanimate materials (displayed as the cosmic abundance of the elements) [3]. Many macroscopic biological features give the appearance of being associated with the scaled repetition of a basic pattern (such as are described by Mandlebrot sets [4] among others).

There is, as yet, no unambiguous evidence of life having existed elsewhere, beyond the Earth, in the Solar System or, indeed, elsewhere in the Universe. The details of the origins of life on Earth are still not known

129

J. S. Shiner (ed.), Entropy and Entropy Generation, 129–137.

precisely except that it is known to have started during the first 0.5Gy
(1 Gy = 1 000 million years) of the Earth's history [5, 6]. Our current
understanding of living processes has led some authors to doubt whether
ordered self-reproducing systems could have formed ab initio during this
time scale [7] but it is most likely that elementary life did form very easily
on the new Earth [8]. The controversy arises because recent astronomical
evidence shows that a wide range of precursor molecules exist widely in
the Universe [9] and would have been available for the formation of early
life on Earth. Equally they have been, and are, available for the formation of living materials elsewhere in the Universe wherever conditions are
favourable. Nevertheless, the question of whether living material is, and
has always been, unique to the Earth remains unresolved. It is difficult to
identify a role played by living materials in the Solar System or in the Universe - it is not even clear that asking for a role is a meaningful question,
although homo sapiens appears to accept that the observed Universe is a
single, logically based entity that can be understood in some sense. Just as
this involves some exploration of an early evolution of the Universe so it
seeks to place living material within some associated ordered context.

The distinction between elementary and advanced life is important. The
first can move to become the second only if the local environment is stable.
The early Solar System was characterised by substantial collisions between
the various bodies and particularly a bombardment of the planets by plan-
etisimals, some quite substantial. The frequency of collision for the ter-
restrial planets (including Earth) was determined in part by the distribution of the major planets (especially Jupiter) whose large masses attracted
many objects away from orbits that would have taken them into the inner
System. Without these dominant planets the frequency of disruptive bombardment of Earth would have been, and would now be, critically greater
than it was / is. This factor is important within the Solar System and would
be one factor influencing the appropriateness of conditions for advanced life
in any other planetary system.

2. Life and Information

It is now generally accepted that biological material is a source of information through the structure of DNA [10]. This information bank ensures the
survival of the present generation, and the formation of the next one: the
evolutionary process can be regarded as the enlargement of the reservoir of
information in the past. Categorising the information content is a complex
problem. A genome contains information for replication both general (the
type of species being replicated) and detailed (whether an arm, leg, wing,
etc). There must also be an action profile to ensure the information is used.

Each genome, therefore, is comprised of a general information store and a specific one. Presumably, the more complicated the species the greater the information store - it is not obvious a priori that homo sapiens represents the maximum information that can be stored this way. The tenacity of living materials to survive is a measure of the stability of an information store once established. The observed evolutionary sequence of living materials can be interpreted as a propensity for the overall information content to increase with time, though not necessarily continuously.

The forces holding this information store together are the forces of interaction between the atoms comprising the DNA strands forming the genome. This has an energy of the order of 10^{-10} J. Mutations of reproduction arise from several causes: one is the reception of energy from outside in excess of that for bonding (for instance action of high energy photons - X-rays, γ-rays and so on); another is the action of a faulty template replication (such as could be associated with irreversible molecular processes common in inanimate matter [11]). Such faults, and especially those of irreversibility, arise during the life of an individual, the template replication particularly deteriorating with time (ageing). Presumably, more complex living forms require more instructions (greater information) for replication than more simple ones, although it has not yet proved possible to express this expectation in quantitative form. Accepting this, the evolutionary sequence can be regarded as a growing concentration of information both as to numbers within a given species and in the growing complexity associated with the evolutionary advancement. The information content for the members of a given species is determined by a variety of environmental factors: the appearance of a new advanced species is an additional information reservoir. The information content of the more elementary life forms will account for only a small proportion of the total information associated with all living materials and can persist. In this way the simultaneous existence of elementary and advanced bodies can be accounted for within the evolutionary manifestation of the "arrow of time". The presence of living material in the Universe therefore represents a store of information that grows with the growing living populations but particularly with the evolutionary development of the complexity of species. In these terms the information content of the Universe can be said to be increasing through the presence of living materials.

3. Cosmology: the Inanimate Universe

The realisation of the vastness of the physical Universe through developments in astronomy [11] has led to the extension of the Newtonian physics (hugely successful in the description of the local everyday Universe) into relativistic physics [12]. The development of general relativity by Einstein through the Principle of Equivalence [13] links the energy / momentum components of matter to a geometric structure constructed from space components and time, providing the basis for a set of field equations to describe the classical large scale behaviour of the Universe. The field equations can be expressed in their simplest form if the structure of the Universe is presumed to be homogeneous and isotropic. Then the geometric metric has the Robertson - Walker form [14] which is the basis of essentially all the current descriptions of the inanimate world on the large scale. The most general solutions of the Robertson - Walker field equations are time dependent.

In the theory the inertia force due to the kinetic motion of the material is in competition with the gravitational force of attraction between the material constituents [15]. The relative importance of the corresponding kinetic energy K of motion and potential energy of attraction, P, characterises the overall behaviour of the Universe and especially its behaviour in time. If $K >> P$ the kinetic energy causes the Universe to expand continuously; if $P >> K$ the gravitational force dominates and the Universe contracts; and if $K = P$ there is a critical case where an initially confined Universe will expand indefinitely or contract indefinitely with increasing speed. The observed Hubble expansion, interpreted as a Doppler effect, indicates an expanding Universe and it is accepted that $K \geq P$.

There is here the possibility of offering a history of the Universe. An expansion of the material Universe from an initial limited size (in space - time) to its present condition is the description offered by the big-bang cosmology [16]. Both space and time are manufactured as a part of the evolution although the material composition is conserved from the beginning but energy is not necessarily conserved locally. Predictions of the theory are the observed background black body micro-wave radiation [17] and the distribution of the elements in the observed cosmic abundance [18]. Alternatively, a time-varying condition is the basis of the theory of the steady state [19]. Here there is a local conservation of energy between material particles and the energy of geometrical stress. It is interesting to notice that none of these descriptions, with the Universe having either a unique origin or not, involves animate materials per se. The arguments often involve what is called an observer but this is a collection of clocks and distance measuring devices which are not animate. Animate materials can be introduced through some anthropic principle [20] which reviews an essentially infinite array of Universes but there is still no attempt to make a unique link between the animate and inanimate worlds. It is also assumed that the living material, where it exists, has consciousness and is intelligent.

The description of the inanimate Universe in these terms involves two further features. First, exotic behaviour of the material universe must be appealed to as part of high energy physics quantum mechanics [21]. Secondly, the behaviour of the vacuum becomes of crucial importance [22]. Indeed, the condition $K = P$ can be interpreted as a vacuum condition in which zero energy is crystallised in kinetic (plus) and potential (minus) forms. This is a generalised application of the approach of d'Alembert [23] to the description of the classical motion of a particle under the action of a force. Although the distribution of luminous matter seems unable to meet this condition, it can well be met by the inclusion of dark matter apparently distributed throughout the Universe [24].

Important daily experiences do not seem to have a place in these various descriptions. There is, for example, no mention of irreversibility. The field equations of Einstein imply adiabatic and isentropic changes [25] and thermodynamic arguments can be incorporated only in their ideal form. Correspondingly, the vacuum does not involve the irreversibility in time [22] which is a central part of everyday experience. This means that the 2nd law of thermodynamics is not applicable in the form of an inequality. There is no increase of entropy in the evolution of the total Universe. The theory describes an ideal world without biological activity and does not, at this level, transfer to the real world of everyday experience.

4. Conservation of Inanimate Quantities

The transfer to the real world is made by realising that the Universe is defined as all that is observed and all that is observable. It follows that nothing lies outside the Universe and that nothing can enter from outside. This means that the details of its evolution must be contained within it and were part of any origin it may have had. In particular, changes of the thermodynamic quantities describing the whole Universe change reversibly and so isentropically. Entropy is conserved for the Universe as a whole as is energy. The behaviour that we recognise as laws again must have its origin within the system and does not derive from outside - this is information through ordering and is to be associated with entropy [26].

The effects of irreversibilities is to increase the entropy locally by an amount $\Delta s(l) > 0$, which is also equivalent to a loss of information. The move to thermodynamic equilibrium is then a move to a condition where there is no information about the initial non-equilibrium state - that is, no further information to lose. The 2nd law of thermodynamics in this context can be accepted as a statement that information is never gained in the inanimate real universe but lost, with the ideal limiting condition that it remains constant.

5. Conservation of Information: An Hypothesis

We now conclude that entropy increases locally (the information content decreasing locally) during the evolution of the inanimate Universe so that $\Delta s(l) > 0$. At the same time, information is gained by the continued evolution of living materials, interpreted as an entropy decrease $\Delta s(b) < 0$. The entropy of the Universe as a whole remains constant so that $\Delta s = \Delta s(l) + \Delta s(b) = 0$. This suggests the general statement $\Delta s(l) = -\Delta s(b)$.

It must be admitted that no unique quantitative way of prescribing the information content of either inanimate or animate matter is yet available so a quantitative comparison between the two changes is not possible. It is interesting, however, to form the hypothesis for the Universe as a whole that

the information content lost by the inanimate world is retained by the evolutionary development of the animate world

This will allow the entropy of the Universe as a whole to remain constant, as is required by thermodynamics for a closed system. The precise replication of members of an individual species, not in thermodynamic equilibrium with the local environment, is a pre-requisite for the maintenance of the information content at a particular time. Biological activity involves irreversible processes. In this way, the inanimate and animate worlds form a single entity which maintains a constant information content for the Universe as a whole.

The conservation of information in both animate and inanimate worlds is achieved through the interaction forces linking the constituent atoms [11]. Various processes, such as the radioactive decay of heavy atoms, the fission of light atoms, the overall spatial expansion of the Universe (expressed through the local Hubble constant [27]) among others, represent a loss of information in real time. Gravitational interactions also represent an information link. Interatomic forces are stronger than gravitational forces at small particle separation distances so atomic aggregation is an effective way of increasing information locally. The observation of quite complex molecules widely distributed in space shows this process does, indeed, occur. A sufficient aggregation of atoms to form large complexes is the requirement for the occurrence of elementary living material. The preferred aggregation will be that leading to the most effective and developing information store and this is the complex of living materials. Normal evolutionary sequences are then to be regarded as the macroscopic manifestation of an increasing information store at the microscopic level. This is to be interpreted as an alternative specification of the uni-direction of time-development portrayed in irreversible thermodynamics. It excludes complicated materials made artificially, including computer developments, since these are not part of a self-growing information sequence. The development of DNA-based living materials on the elementary RNA bases is fully consistent with our hypothesis.

The above hypothesis places animate and inanimate materials on a comparable footing in the Universe - both evolve in such a way that the total information content is conserved. A fundamental role is also provided for the presence of irreversible processes in the real world.

The development of animate/inanimate material obviously requires sufficient inanimate form to provide an environment for the animate form. The increasing provision of heavier chemical atoms by stellar evolution acts as a source for future living materials, the quantity and complexity of such materials increasing with time. It would follow that, in so far as there is a limit to the provision of inanimate materials, there is an ultimate limit to the complexity of life forms.

The hypothesis requires detailed testing in the future. Of especial importance is the quantitative specification of the information content of DNA and the further elucidation of the relation between entropy and order in the inanimate world.

6. Summary and Conclusions

An attempt is made to provide a logical place within the cosmos for living materials which have the characteristic of defying the second law of thermodynamics in not attaining an equilibrium with the immediate environment. Individual members of a species achieve this for limited amounts of time but the continual replenishment of the species as a whole maintains this dis-equilibrium. This is at root a discontinuous process (a "life / death" balance) but the fluctuations are small. As a closed system, the Universe (encompassed in our event horizon) must evolve at constant entropy but irreversible inanimate processes cause an increase of entropy locally. It is proposed, through an hypothesis, that animate materials provide the negative entropy to maintain that for the Universe at a constant value everywhere locally through the associated information content. Classical evolution is then to be regarded as the manifestation, at the everyday level, of the continual increase of information at the molecular level.

It follows from this point of view that the development of living materials is a strictly local phenomenon everywhere. This also implies that living materials are not in principle unique to the Earth. While every locality is a potential centre for living materials, only those localities able to support the material development will show life. The development at a particular location is independent of developments elsewhere, and this isolation is consistent with the known scale of the Universe. The living material will tend

to fill its environment if it is established with sufficient possibilities to develop intelligence to an appropriate level. This draws a distinction between
the formation of living materials per se and the development of intelligent
materials as a secondary consideration.

These consequences fit well with experience. Life on Earth is set to fill
the Solar System given sufficient time (although it is not clear in what way
this will influence the expected evolution of the System, if at all). The lack
of any evidence of life forms elsewhere in the Universe (sought through the
SETI programme) is fully consistent with this, and with the arguments
giving "Drake's equation" [28, 29]. It is reasonable to conclude that the
level of complexity of living material at any location is determined by the
age of the local environment so a spread of complexities throughout the
Universe can be expected even though communication between different
forms is not necessary for the separate evolution of each. The limits of any
physical communication between different intelligent centres is likely to be
set by the average lifespan of the individual members [30]. There is no way
one can ascertain the magnitude of this lifespan but it would need to be
measured in thousands of Earth years if physical communication (such as
visitations) were to be achieved generally.

The testing and further development of these ideas must involve the
quantification of the ordering within living materials, the elucidation of the
importance of chirality [31] and, above all, definitive astronomical evidence
for or against the presence of living materials elsewhere in the Universe. The
early surface histories of Venus and Mars [32] are such as not to rule out
the possibilities of primitive life having formed there early in the history of
the Solar System. The possibility now of primitive life within the icy satel-
lites of Jupiter and Saturn, comparable to the living materials found near
"smokers" of the mid-ocean ridges on Earth, cannot be definitely excluded
[33].

Acknowledgements

I am pleased to thank Professor Sir William McCrea, FRS, of the University of Sussex, Dr. N Brindle of the University of Leicester and Dr. M.
Pasenkiewicz-Gierula of the Jagiellonian University, Krakow, for very helpful comments on an earlier draft of the manuscript.

References

1. Geenberg, J.M., Mendoza-Gomez, C.X. and Pirronello, V. (eds.) (1993) *The Chem-
    istry of Life's Origins* (NATO ASI Series), Kluwer, Dordrecht.
2. Gould, S.J. (1995) *Life in the Universe, Scientific American Special.*
3. Cameron, A.G.W (1973) *Space Sci. Rev.* **15**, 121- 146; Anders, E. and Grevesse, N.
    (1989) *Geochim. Cosochim. Acta*, **46**, 289-298.

4.   Mandlebrot, B.B. (1982) *The Fractal Geometry of Nature*, W.H. Freeman, Oxford; Devlin, K. (1994) *Mathematics: The Science of Patterns*, Scientific American Library.
5.   Schof, J.W. and Packer, B.N. (1987) *Science*, **237**, 70-73.
6.   Plug, H.D. (1978) *Naturwissenschaften*, **65**, 611-615.
7.   Hoyle, F. and Wickramasinghe, N.C. (1986) *Quart. J. Roy. Ast. Soc.*, **27**, 21-37.
8.   Ferris, J.P. (1995) *Nature*, **373**, 659.
9.   see e.g. Williams, D.A. (1986) *Quart. J. Ast. Soc.*, **27**, 64-70; Duley, W.W. (1986) *Quart. J. Ast. Soc.*, **27**, 403-412.
10. Szathmary, E. and Maynard Smith , J. (1995) *Nature*, **374**, 227-232.
11. see e.g. Shu, F.H. (1982) *The Physical Universe*, Oxford University Press, Oxford; Zubarev, D.N. (1973) *Nonequilibrium Statistical Thermodynamics*, Consultants Bureau, London.
12. McCrea, W. (1960) *Relativity Physics*, Methuen Monographs on Physical Subjects, London.
13. Tolman, R.C. (1934) *Relativity, Thermodynamcis and Cosmology*, Clarendon Press, Oxford.
14. Sciama, D.W. (1975) *Modern Cosmology*, Cambridge University Press, Cambridge.
15. see e.g. Bondi, H. (1952) *Cosmology*, Cambridge University Press, Cambridge.
16. Hawking, S.W. and Israel, W. (eds.) (1979) *General Relativity*, Clarendon Press, Oxford.
17. see ref. (14), Chap. 14.
18. Tayler, R.J. (1990) *Quart. J. Ast. Soc.*, **31**, 281-304.
19. see ref (15), Chap. XII.
20. see e.g. Sagan, C. and Newman, W.I. (1983) *Quart. J. Ast. Soc.*, **24**, 113-121; Harrison, E.R. (1995) *Quart. J. Ast. Soc.*, **36**, 193-203.
21. Reece, Sir. M. (1993) *Quart. J. Ast. Soc.*, **34**, 279- 290; Weinberg, S. (1977) *The First Three Minutes*, Andre Duetsch, London.
22. McCrea, Sir W. (1986) *Quart. J. Ast. Soc.*, **27**, 137- 152.
23. see e.g. Lanczos, C. (1949) *The Variational Principles of Mechanics*, University of Toronto Press, Toronto.
24. see Sciama, D.W. (1993) *Quart. J. Ast. Soc.*, **34**, 291-300.
25. Lavenda, B. H. and Dunning-Davies, J. (1992) *Found. of Phys. Lett.*, **5**, 191-196.
26. Prigogine, I and Stengers, I. (1985) *Order Out of Chaos*, Fontane Paperback, London.
27. see ref (15).
28. Drake, F.D. Discussion at Space Science Board of National Academy of Sciences, Washington, D.C.; Lytkin, V., Finney, B. and Alepko, L. (1995), *Quart. J. Ast. Soc.*, **36**, 369-376.
29. Bates, D.R. (1988) *Quart. J. Ast. Soc.*, **29**, 307-312.
30. Cole, G.H.A. (1996) *Quart. J. Ast. Soc.* (in the press).
31. Bonner, W.E. (1991) *Origs. Life and Evol.Biosph.*, **11**, 59-64.
32. Briggs, G.A. and Taylor, F.W. (1982) *Photographic Atlas of the Planets*, Cambridge University Press, Cambridge.
33. Cole, G.H.A (1988) *Phil. Trans. Roy. Soc. London*, **A235**, 569-582.

PHENOMENOLOGICAL MACROSCOPIC SYMMETRY IN DISSIPATIVE NONLINEAR SYSTEMS

J.S. SHINER

Physiologisches Institut, Universität Bern
Bühlplatz 5, CH-3012 Bern, Switzerland

AND

STANISLAW SIENIUTYCZ

Department of Chemical Engineering, Warsaw Technical
University, 1 Warynskiego Street, 00-645 Warsaw, Poland

Abstract.

We show that systems whose equations of motion are derivable from a Lagrangian formulation extended to allow for dissipative effects by the inclusion of dissipative forces of the form (resistance X flow) and subject to constraints expressing conservation of quantities such as mass and charge possess a symmetry property at stationary states arbitrarily far from thermodynamic equilibrium. Examples of such systems are electrical networks, discrete mechanical systems and systems of chemical reactions. The symmetry is in general only an algebraic one, not the differential Onsager reciprocity valid close to equilibrium; the more general property does reduce to the Onsager result in the equilibrium limit, however. No restrictions are placed on the form of the resistances for dissipative processes; they may have arbitrary dependencies on the state of the system and/or its flows. Thus, the symmetry property is not limited to linear systems. The approach presented here leads naturally to the proper set of flows and forces for which the Onsager-like symmetry holds.

1. Introduction

Symmetry relations play a fundamental role in our understanding of nature [1]. Not least among these is the symmetry displayed by the phenomenolo-

139

J. S. Shiner (ed.), Entropy and Entropy Generation, 139–158.
© 1996 *Kluwer Academic Publishers.*

gical equations of near equilibrium thermodynamics [2, 3]:

$$X_i = \sum_j \mathcal{R}_{ij} J_j \tag{1}$$

Here the X_i and the J_j are the phenomenological forces and fluxes, respectively, and $\mathcal{R}_{ij}$ are known as phenomenological resistances. The symmetry property, also known as Onsager reciprocity, is

$$\mathcal{R}_{ij} = \mathcal{R}_{ji} \tag{2}$$

or, since the phenomenological resistances are constants evaluated at thermodynamic equilibrium

$$\partial X_i / \partial J_j = \partial X_j / \partial J_i \tag{3}$$

Onsager reciprocity is a macroscopic symmetry founded on the microscopic property of time reversal symmetry and is based on the study of fluctuations around equilibrium [1, 2, 3, 4, 5]. In general it is an equilibrium property based on microscopic reversibility or detailed balance and is valid only in the near equilibrium regime, which we define by the condition that the phenomenological equations (1) are valid with constant phenomenological resistances $\mathcal{R}_{ij}$:

$$X_i = \sum_j \mathcal{R}_{ij} J_j, \ \mathcal{R}_{ij} \ constant \tag{4}$$

Because of the significance of symmetry in general and especially for nonequilibrium thermodynamics there have been many attempts to extend the phenomenological equations and Onsager reciprocity to far from equilibrium regimes, or at least to find such regimes where Onsager reciprocity is valid [6, 7, 8], but when such regimes have been found they have shown themselves to be special cases, and in general Onsager reciprocity is valid only in the near equilibrium regime. Keizer [5, 9, 10] among others [11, 12, 13] has shown that equations of the form of eq. (1) can be derived for stable stationary states far from equilibrium if the phenomenological forces and fluxes are replaced by their deviations from their stationary state values. These derivations are based on a Taylor expansion to first order around the stationary state and, like those of Onsager for the near equilibrium case, are based on the study of fluctuations. Nonetheless, reciprocity is not found except for the near equilibrium regime.

For some systems the range of validity of the phenomenological relations and Onsager reciprocity extends surprisingly far from equilibrium [1]. Examples would be electrical networks with constant resistances and discrete mechanical systems with constant friction coefficients. However, if the behavior of the resistances and friction coefficients is not ideal, i.e. if they are not constant, then the validity of near equilibrium thermodynamics is restricted to a domain much closer to equilibrium. For other sorts of processes, e.g. chemical reactions, the relation between conjugate thermodynamics forces and fluxes is inherently nonlinear, even in the ideal case. For these processes the near equilibrium regime is strictly constrained to the condition given in eq. (4).

Taking a more macroscopic phenomenological approach, one of us has recently shown [14, 15] that the stationary state dynamics of systems of chemical reactions arbitrarily far from thermodynamic equilibrium can be written in the form of the phenomenological equations of nonequilibrium thermodynamics [eq. (1)] and that the symmetry property of eq. (2) is also valid arbitrarily far from equilibrium. This approach is not based on a linear expansion around the stationary state nor on the study of fluctuations; rather it contains the full, possibly nonlinear dynamics of the system. Unlike the case for the original near equilibrium phenomenological equations the phenomenological resistances $\mathcal{R}_{ij}$ for this macroscopic phenomenological approach are not in general constant but may depend on the state of the system and its fluxes. Furthermore, the symmetry is in general only algebraic, not differential as is the case for Onsager near equilibrium reciprocity; i.e., the symmetry property of eq. (3) does not hold in general.

We will show here that the algebraic symmetry property derived from the macroscopic phenomenological approach is not restricted to systems of chemical reactions but is valid for a general class of systems having the following properties: (1) the equations of motion can be derived from the Lagrangian formalism extended to nonconservative systems by the inclusion of dissipative force terms [15, 16, 17, 18]; (2) the dissipative forces can be written in the form $R\dot{q}$, where R is the resistance of the dissipative process and $\dot{q}$ its rate or flow; (3) for each state variable there is a conservation relation relating its change to the dissipative flows in the system and the flows through external sources, of which there may be one controlling each state variable; these conservation relations take a form analogous to Kirchhoff's current law for electrical circuits. Systems possessing these three properties may include the systems of chemical reactions already treated, electrical networks and discrete mechanical systems; additionally, heterogeneous systems involving processes of these types – electrochemical, mechanochemical and electromechanical systems – may also possess these properties. Some examples are given in the figures below. We exclude for the time being systems with magnetic properties [19]; we will perhaps return to the question of the effects of magnetic fields in a future publication. After proving the symmetry property for system with these properties we will discuss its implications and its relation to Onsager reciprocity.

2. Definitions of Systems Considered: Variables, Forces, Constraints and Equations of Motion

We distinguish between two different classes of variables in the systems considered here, with one of the classes divided into two subclasses. Variables of the first class we call state variables and are divided into internal state variables and (external) source state variables. Examples of internal state variables would be the charges on capacitors in electrical networks, lengths in discrete mechanical systems or mole numbers in a system of chemical reactions. Examples of source state variables would be the charges conjugate to ideal potential sources in electrical networks and the lengths conjugate to the external forces in discrete mechanical systems. Examples may be seen in Figs. 1 and 2. We use the symbol q_{s_i} for the ith internal state variable and q_{x_i} for the ith source state variable. Variables of the second class are called dissipative process variables; we use the symbol q_{p_j} for the jth process variable. The time derivatives $\dot{q}_{p_j}$ of these variables are the rates of the dissipative processes occurring in the system. Examples of these process rates are currents through resistors in electrical networks, rates of chemical reactions and mechanical velocities through dashpots (Figs. 1 and 2). We let N_s, N_x and N_p be the total numbers of internal state variables, external sources and dissipative processes, respectively. The meaning of these various classes of variables will become clear in the sections below, if it is not so already. [Note that the distinction between the variables q_{s_i}, q_{x_i} and q_{p_j} by no means implies that they are independent variables; in fact they are related by the constraints expressed in eq. (16) below.]

For the systems under consideration here the equations of motion may be written in terms of a dissipative Lagrangian formulation [16, 17, 18, 20] as

$$\frac{d}{dt}\frac{\partial \mathcal{L}}{\partial \dot{q}} - \frac{\partial \mathcal{L}}{\partial q} + \frac{\partial \mathcal{F}}{\partial \dot{q}}\bigg|_R + \sum_l \lambda_l \frac{\partial f_l}{\partial \dot{q}} = 0 \qquad (5)$$

where the q is one of the internal state, external source or dissipative processes variables. $\mathcal{L}$ is the Lagrangian, $\mathcal{F}$ is a generalized Rayleigh dissipation function, the f_l are equations of constraint, and the λ_l are undetermined multipliers. The notation in the third term on the left hand side should indicate that the resistances R of the dissipative processes are held constant when taking the partial derivative of the dissipation function with respect to the time derivative of the variable q. We now proceed to explain the various terms in the equation of motion.

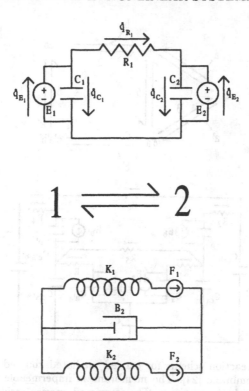

Figure 1. A simple electrical circuit (top), a simple chemical reaction (middle) and a simple discrete mechanical system (bottom) with analogous properties. Not shown for the chemical reaction are the external sources for species 1 and 2 which are analogous to the external sources E_i in the electrical circuit and the external forces F_i in the mechanical system.

2.1. LAGRANGIAN FORCES

The Lagrangian $\mathcal{L}$ is defined as

$$\mathcal{L}\left(\overline{q}_s, \dot{\overline{q}}_s, \overline{q}_x, \dot{\overline{q}}_x\right) = T^{\bullet} - U \tag{6}$$

where $T^{\bullet}$ is the kinetic coenergy, U is the potential energy (or the appropriate Legendre transformation thereof), $\overline{q}_s$ is the vector of internal state variables and $\overline{q}_x$ is the vector of external source variables. Note that we have written $\mathcal{L}$ as independent of the dissipative process variables and their rates. This is consistent with our classification of variables as either state variables or dissipative process variables; since the Lagrangian as defined here contains no dissipative effects, we should be able to write it as independent of the dissipative process variables and rates.

For the electrical example in Fig. 1 the internal state of the system is determined simply by the charges on the two capacitors; therefore the internal state variables are q_1 and q_2. There are two external sources in this example,

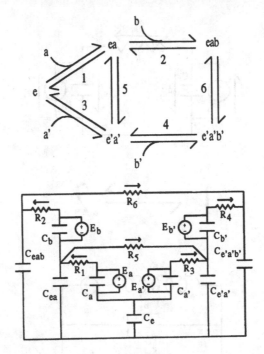

Figure 2. Top. A reaction scheme for carried mediated coupled transport of species a and b across a membrane [24]. The membrane is impermeable to a and b but not to their complexes with the carrier e. The unprimed species reside on one side of the membrane and the primed species on the other side. The chemical potentials of species a and b on both sides of the membrane are controlled by external sources analogous to ideal potential sources in electrical circuits. Bottom. An electrical circuit analogous to the reaction scheme (top).

E_1 and E_2. The flows (electrical currents in this case) through these sources are $\dot{q}_{E_1}$ and $\dot{q}_{E_2}$ respectively. The corresponding external sources variables are thus q_{E_1} and q_{E_2}. The Lagrangian may thus be written as

$$\mathcal{L} = -\left(q_1^2/C_1 + q_2^2/C_2 - E_1 q_{E_1} - E_2 q_{E_2}\right) \qquad (7)$$

Note that the contributions to the Lagrangian involving the internal state variables and those involving the external sources have opposite sign. This sign difference arises from our choice of the network convention [20], where current is assumed to flow into the positive pole of a passive network element but out of the positive pole of a source element. For the mechanical example the Lagrangian takes an analogous form with lengths replacing charges and external forces replacing potential sources.

For the chemical example in Fig. 1 the Lagrangian is

$$\mathcal{L} = -\left(n_1 \mu_1 + n_2 \mu_2 - n_1^x \mu_1^x - n_2^x \mu_2^x\right) \qquad (8)$$

where the n_i are mole numbers, the μ_i chemical potentials and superscript 'x' denotes the sources. The source μ_i^x holds the chemical potential μ_i at the value of the chemical potential of the source, just as the electrical sources in the electrical example determine the electrical potential drops over the capacitors. The $\dot{n}_i^x$ are mass flows through these chemical sources, analogous to current flows through electrical sources. Ross et al. [21] have discussed how such chemical sources may be realized.

All these examples fall under a general formulation in which the generalized internal forces conjugate to the internal state variables are given by

$$\phi_i = \frac{d}{dt}\frac{\partial \mathcal{L}}{\partial \dot{q}_{s_i}} - \frac{\partial \mathcal{L}}{\partial q_{s_i}} \tag{9}$$

If we assume that the capacitances of the electrical examples are constant, then these internal forces are just $-\partial \mathcal{L}/\partial q_1 = q_1/C_1$ and $-\partial \mathcal{L}/\partial q_2 = q_2/C_2$. For the chemical example the internal forces are $-\partial \mathcal{L}/\partial q_1 = \mu_1$ and $-\partial \mathcal{L}/\partial q_2 = \mu_2$. If one would wish to define chemical capacitances by analogy to the result for the electrical case, $-\partial^2 \mathcal{L}/\partial q_i^2 = 1/C_i$, the inverse chemical capacitances would then be given by $-\partial^2 \mathcal{L}/\partial n_i^2 = \partial \mu_i/\partial n_i$. This analogy is not too important, however, particularly since we do not assume constant electrical capacitances in general. What is important is that these terms in the Lagrangian represent passive energy storage and inertial effects, i.e. the ϕ_i are non- dissipative forces. Similarly, the generalized external forces conjugate to the source state variables are given by

$$\phi_i^x = \frac{d}{dt}\frac{\partial \mathcal{L}}{\partial \dot{q}_{x_i}} - \frac{\partial \mathcal{L}}{\partial q_{x_i}} \tag{10}$$

For the electrical and mechanical examples the external forces are just the electrical potential differences of the electrical sources and the mechanical external forces, respectively. The sources for systems of chemical reactions have been discussed previously [14, 15]; their effect in the example of Fig. 1 would be to control the chemical potentials of the species 1 and 2. In general it is necessary to allow for such sources if a system is to achieve nonequilibrium stationary states, whether we are dealing with chemical, electrical or mechanical systems. More involved examples of a system of chemical reactions and an electrical circuit with analogous properties are given in Fig. 2.

2.2. DISSIPATIVE FORCES

The dissipative forces are derivable from the generalized Rayleigh dissipation function, which may be written

$$\mathcal{F} = \frac{1}{2} \sum_{j=1}^{N_p} R_j \left(\overline{q}_s, \dot{\overline{q}}_s, \overline{q}_x, \dot{\overline{q}}_x, \overline{q}_p, \dot{\overline{q}}_p \right) \dot{q}_{p_j}^2 \qquad (11)$$

where R_j is the resistance of the jth dissipative process. Note that only the rates of the dissipative processes $\dot{q}_{p_j}$ appear explicitly in the sum, although the resistances may depend on any of the $\dot{q}_{s_i}$ and $\dot{q}_{x_i}$ as well. They may also depend on any of the q_{p_j}, q_{s_i}, and q_{x_i}. The third term on the left of the equations of motion (5) is the dissipative force conjugate to the disspative rate $\dot{q}_{p_j}$ and is thus

$$\left. \frac{\partial \mathcal{F}}{\partial \dot{q}_{p_j}} \right|_R = R_j \dot{q}_{p_j} \qquad (12)$$

since the resistances R_j are held constant when the partial differentiation is performed. This force has the general form for a dissipative process:

force driving a dissipative process
= resistance of the process X rate of the process.

This form is well known from electrical networks, where the potential difference across a resistor = the resistance X the current through the resistor, and from mechanics, where the frictional force can be written as the product of a friction coefficient and the velocity. Chemical reaction mass action kinetics can also be written in this form, where the force is the affinity of the reaction and the rate of the process is simply the rate of the reaction, when the "chemical" resistance is properly defined [14, 15, 22, 23]. Examples of these dissipative forces are

$$R_1 \dot{q}_{R_1}, \ R_2 \dot{q}_{R_2}, \ \cdots$$

for the electrical network of Fig. 2 and

$$R_{ch_1} \dot{\xi}_1, \ R_{ch_2} \dot{\xi}_2, \ \cdots$$

for the system of chemical reactions. $\dot{q}_{R_j}$ is the current through the resistor R_j, R_{chj} is the chemical resistance of the jth reaction, and $\dot{\xi}_j$ is the rate of reaction j. The corresponding dissipation functions are

$$\mathcal{F} = \frac{1}{2} \sum_{j=1}^{6} R_j \dot{q}_{R_j}^2 \qquad (13)$$

for the electrical example, and

$$\mathcal{F} = \frac{1}{2} \sum_{j=1}^{6} R_{ch_j} \dot{\xi}_j^2 \tag{14}$$

for the system of chemical reactions.

We should emphasize here that no restrictions are placed on the resistances, whether they be electrical, mechanical, chemical or whatever. They may in general depend on all the variables of the system as well as their time derivatives:

$$R_j \left(\bar{q}_s, \dot{\bar{q}}_s, \bar{q}_x, \dot{\bar{q}}_x, \bar{q}_p, \dot{\bar{q}}_p \right)$$

Indeed the form of the chemical resistance derived previously [14, 15, 22, 23] is a nonlinear function of the q_{s_i}. It could also be written as a function of the q_{s_i} and the rate of the reaction There is also no requirement that electrical resistances or mechanical friction coefficients be constant; we allow them to be any function of the charges and currents in the network in the former case and any function of lengths and velocities in the latter. Whittaker [18] has noted that Newton and J. Bernoulli considered mechanical resistive forces dependent on higher orders of velocity as early as three centuries ago.

Nonetheless, when the resistances R_j of the dissipative processes are not constant, the condition that the R_j are to be held constant when partial derivatives of the dissipation function are taken with respect to the rates of the dissipative processes and other velocities in the equations of motion (eq. 5) may seem somewhat arbitrary. However, as noted by Rayleigh [16], the resistances need not be constant for the quadratic form of the dissipation function to hold; it is only necessary that they be independent of the rates of the dissipative processes. This is the case for the chemical resistances determined previously [14, 15, 22, 23]. Moreover, as we have recently shown in the context of systems of chemical reactions [25, 26], a more general formulation shows that the quadratic form of the Rayleigh dissipation function may be valid even when the resistances depend on the rates of the dissipative processes or other velocities. In this general formulation the equations of motion are determined by the condition that the power criterion

$$P \equiv \frac{d\mathcal{G}}{dt} + \mathcal{F} + \Psi + \frac{\partial \mathcal{L}}{\partial t} \tag{15}$$

be an extremum, where

$$\mathcal{G} \equiv \sum_i \dot{q}_i \frac{\partial \mathcal{L}}{\partial \dot{q}_i} - \mathcal{L}$$

is the Legendre transform of the Lagrangian with respect to the velocities in the system, $\mathcal{F}$ is the generalized Rayleigh dissipation function as given above, and

$$\Psi \equiv \frac{1}{2} \sum_i \left(\frac{d}{dt} \frac{\partial \mathcal{L}}{\partial \dot{q}_i} - \frac{\partial \mathcal{L}}{\partial q_i} \right)^2 / R_i$$

is the dissipation function of the second kind. The equations of motion determined by this extremum principle yield dissipative forces of the form $R_i \dot{q}_i$ regardless of any dependence of the resistances on the q_i, the $\dot{q}_i$ or even the accelerations in the system. Thus the prescription of holding the resistances constant when taking the partial derivatives of the dissipation function with respect to the rate of the dissipative processes in the dissipative Lagrangian equations (5) leads to the correct equations of motion. [It should be noted that the general formulation of eq. (15) requires that one choose the q_i to be an independent set of variables.]

2.3. CONSERVATION PROPERTIES

The rates of change of the state variables and the rates of the processes are related by conservation equations of the form

$$f_i = \dot{q}_{s_i} - \sum_{j=1}^{N_p} \nu_{ij} \dot{q}_{p_j} + \dot{q}_{x_i} = 0, \ 1 \le i \le N_x$$

$$f_i = \dot{q}_{s_i} - \sum_{j=1}^{N_p} \nu_{ij} \dot{q}_{p_j} = 0, \ N_x + 1 \le i \le N_s$$

(16)

where there are a total of N_s state variables, the first N_x of which are controlled by external sources. The ν_{ij} are constant. For a system of chemical reactions they are just the stoichiometric coefficients. For the chemical example in Figure 2 we have

$$
\begin{pmatrix} \dot{n}_a \\ \dot{n}_b \\ \dot{n}_{a'} \\ \dot{n}_{b'} \\ \dot{n}_e \\ \dot{n}_{ea} \\ \dot{n}_{e'a'} \\ \dot{n}_{eab} \\ \dot{n}_{e'a'b'} \end{pmatrix}
-
\begin{pmatrix}
-1 & 0 & 0 & 0 & 0 & 0 \\
0 & -1 & 0 & 0 & 0 & 0 \\
0 & 0 & -1 & 0 & 0 & 0 \\
0 & 0 & 0 & -1 & 0 & 0 \\
-1 & 0 & -1 & 0 & 0 & 0 \\
1 & -1 & 0 & 0 & -1 & 0 \\
0 & 0 & 1 & -1 & 1 & 0 \\
0 & 1 & 0 & 0 & 0 & -1 \\
0 & 0 & 0 & 1 & 0 & 1
\end{pmatrix}
\begin{pmatrix} \dot{\xi}_1 \\ \dot{\xi}_2 \\ \dot{\xi}_3 \\ \dot{\xi}_4 \\ \dot{\xi}_5 \\ \dot{\xi}_6 \end{pmatrix}
-
\begin{pmatrix} \dot{n}_a^x \\ \dot{n}_b^x \\ \dot{n}_{a'}^x \\ \dot{n}_{b'}^x \\ 0 \\ 0 \\ 0 \\ 0 \\ 0 \end{pmatrix}
= \bar{0} \ (17)
$$

This expression relates the rates of change of the mole numbers, the $\dot{n}_i$, to the rates of the reactions, the $\dot{\xi}_j$. Species a and b on both sides of the membrane are controlled by external sources; the flow through these sources, $\dot{n}_i^x$, are given in the last vector on the right hand side. Note that for those species which are not controlled by sources zeros are entered into this vector.

The role of these source terms may be made more clear by examination of the electrical network example of Fig. 2. Here equations (16) take the form

$$
\begin{pmatrix} \dot{q}_{C_1} \\ \dot{q}_{C_2} \\ \dot{q}_{C_3} \\ \dot{q}_{C_4} \\ \dot{q}_{C_5} \\ \dot{q}_{C_6} \\ \dot{q}_{C_7} \\ \dot{q}_{C_8} \\ \dot{q}_{C_9} \end{pmatrix} - \begin{pmatrix} -1 & 0 & 0 & 0 & 0 & 0 \\ 0 & -1 & 0 & 0 & 0 & 0 \\ 0 & 0 & -1 & 0 & 0 & 0 \\ 0 & 0 & 0 & -1 & 0 & 0 \\ -1 & 0 & -1 & 0 & 0 & 0 \\ 1 & -1 & 0 & 0 & -1 & 0 \\ 0 & 0 & 1 & -1 & 1 & 0 \\ 0 & 1 & 0 & 0 & 0 & -1 \\ 0 & 0 & 0 & 1 & 0 & 1 \end{pmatrix} \begin{pmatrix} \dot{q}_{R_1} \\ \dot{q}_{R_2} \\ \dot{q}_{R_3} \\ \dot{q}_{R_4} \\ \dot{q}_{R_5} \\ \dot{q}_{R_6} \end{pmatrix} - \begin{pmatrix} \dot{q}_{E_1} \\ \dot{q}_{E_2} \\ \dot{q}_{E_3} \\ \dot{q}_{E_4} \\ 0 \\ 0 \\ 0 \\ 0 \\ 0 \end{pmatrix} = \bar{0} \quad (18)
$$

Notice that these are identically analogous to those for the chemical reaction scheme of Fig. 2. In fact, the network was chosen for this analogy. The q_C's are the charges on the capacitors and are the internal state variables; the q_E's are the charges conjugate to the ideal potential sources and are the source state variables; the $\dot{q}_R$'s are the currents through the resistors and thus the process rates for the network. The equations of constraint (18) are just Kirchhoff's current law at the capacitors and express conservation of charge. For the chemical example the analogous relations (17) express conservation of mass, one equation for each species involved in the reaction scheme.

Since the analogy between discrete mechanical systems and electrical networks is well known [17, 20], we simply state here that the example of Fig.2 could also be interpreted as a mechanical system and that all results presented here apply also to discrete mechanical systems.

2.4. THE EQUATIONS OF MOTION

From equations (5, 9-11) the equations of motion for the systems under consideration are now given by

$$
\begin{aligned}
\phi_i + \lambda_i &= 0, \ 1 \leq i \leq N_s \\
\phi_i^x - \lambda_i &= 0, \ 1 \leq i \leq N_x \\
R_j \dot{q}_{p_j} - \sum_{i=1}^{N_s} \nu_{ij} \lambda_i &= 0, \ 1 \leq i \leq N_p
\end{aligned} \quad (19)
$$

or, upon eliminating the λ_i,

$$
\begin{aligned}
\phi_i &= -\phi_i^x, \ 1 \leq i \leq N_x \\
R_j \dot{q}_{p_j} &= -\sum_{i=1}^{N_s} \nu_{ij} \phi_i = 0, \ 1 \leq i \leq N_p
\end{aligned} \quad (20)
$$

The first of these equations expresses the control of the force ϕ_i conjugate to the ith internal state variable by the corresponding external force ϕ_i^x for the first N_x internal variables, and the second is the relation between the rate of the jth process and its conjugate driving force. Note that the forces ϕ_i and ϕ_i^x may be nonlinear.

3. Independent and Dependent Variables and The Stationary State

We first write the relation between the rate of the processes and their conjugate forces (eq. 20) and the conservation relations involving the rates of change of the state variables and the rates of the processes (eq. 16) in vector form:

$$R\dot{\bar{q}}_p = -\nu^T \bar{\phi}$$

$$\tag{21}$$

$$\dot{\bar{q}}_s = \nu \dot{\bar{q}}_p + \begin{pmatrix} \dot{\bar{q}}_x \\ 0 \end{pmatrix}$$

The second vector on the right hand side of the second equation has the elements $\dot{q}_{x_i}$ for $1 \leq i \leq N_x$ and 0 for $N_x + 1 \leq i \leq N_s$. R is the diagonal matrix with elements $R_{ij} = \delta_{ij}R_i$ and ν is the $(N_s \times N_p)$ matrix with elements ν_{ij}. $\bar{\phi}$ is the vector with elements ϕ_i. At the stationary state all elements of $\dot{\bar{q}}_s$ vanish by definition, and the conservation equations become

$$\nu \dot{\bar{q}}_p = - \begin{pmatrix} \dot{\bar{q}}_x \\ 0 \end{pmatrix}$$

$$\tag{22}$$

Thus, all of the $\dot{q}_{p_j}$ are not independent at the stationary state. We arbitrarily choose the last $(N_s - N_x)$ of the $\dot{q}_{p_j}$ to be the dependent rates of the processes and partition the vector $\dot{\bar{q}}_p$ into a vector of independent rates $\dot{\bar{q}}_{pI}$ and a vector of dependent rates $\dot{\bar{q}}_{pD}$:

$$\dot{\bar{q}}_p = \begin{pmatrix} \dot{\bar{q}}_{pI} \\ \dot{\bar{q}}_{pD} \end{pmatrix}$$

$$\tag{23}$$

where

$$\dot{\bar{q}}_{pI} = \left(\dot{q}_{p_1}, tq_{p_2}, \cdots, \dot{q}_{pN_p - N_s + N_x} \right)^T$$

$$\dot{\bar{q}}_{pD} = \left(\dot{q}_{pN_p - N_s + N_x + 1}, \dot{q}_{pN_p - N_s + N_x + 2}, \cdots, \dot{q}_{pN_p} \right)^T$$

Under certain conditions all of the independent rates so defined may not actually be independent [14, 15]. If this is the case we take a smaller set of rates as independent. Since the arguments below remain the same whether all the rates taken as independent in eqs. (23) are truly independent or not, we will assume in this paper that they are truly independent.

We also partition ν as

$$\nu = \begin{pmatrix} \nu_{II} & \nu_{ID} \\ \nu_{DI} & \nu_{DD} \end{pmatrix} \tag{24}$$

where

$$\nu_{II} = \begin{pmatrix} \nu_{11} & \cdots & \nu_{1,N_p-N_s+N_x} \\ \vdots & & \vdots \\ \nu_{N_x\,1} & \cdots & \nu_{N_x,N_p-N_s+N_x} \end{pmatrix}$$

$$\nu_{ID} = \begin{pmatrix} \nu_{1,N_p-N_s+N_x+1} & \cdots & \nu_{1\,N_p} \\ \vdots & & \vdots \\ \nu_{N_x,N_p-N_s+N_x+1} & \cdots & \nu_{N_{sc}\,N_p} \end{pmatrix}$$

$$\tag{25}$$

$$\nu_{DI} = \begin{pmatrix} \nu_{N_x+1,1} & \cdots & \nu_{N_x+1,N_p-N_s+N_x} \\ \vdots & & \vdots \\ \nu_{N_s\,1} & \cdots & \nu_{N_s,N_p-N_s+N_x} \end{pmatrix}$$

$$\nu_{DD} = \begin{pmatrix} \nu_{N_x+1,N_p-N_s+N_x+1} & \cdots & \nu_{N_x+1,N_p} \\ \vdots & & \vdots \\ \nu_{N_s,N_p-N_s+N_x+1} & \cdots & \nu_{N_s\,N_p} \end{pmatrix}$$

The stationary state condition can now be written

$$\begin{pmatrix} \nu_{II} & \nu_{ID} \\ \nu_{DI} & \nu_{DD} \end{pmatrix} \begin{pmatrix} \dot{\bar{q}}_{pI} \\ \dot{\bar{q}}_{pD} \end{pmatrix} = - \begin{pmatrix} \dot{\bar{q}}_x \\ 0 \end{pmatrix}$$

or

$$\tag{26}$$

$$\nu_{II}\dot{\bar{q}}_{pI} + \nu_{ID}\dot{\bar{q}}_{pD} = -\dot{\bar{q}}_x$$
$$\nu_{DI}\dot{\bar{q}}_{pI} + \nu_{DD}\dot{\bar{q}}_{pD} = \bar{0}$$

We can now partition the relation between the rate of the processes and their conjugate forces:

$$\begin{pmatrix} R_{II} & R_{ID} \\ R_{DI} & R_{DD} \end{pmatrix} \begin{pmatrix} \dot{\bar{q}}_{pI} \\ \dot{\bar{q}}_{pD} \end{pmatrix} = \begin{pmatrix} R_{II} & 0 \\ 0 & R_{DD} \end{pmatrix} \begin{pmatrix} \dot{\bar{q}}_{pI} \\ \dot{\bar{q}}_{pD} \end{pmatrix} = - \begin{pmatrix} \nu_{II}^T & \nu_{DI}^T \\ \nu_{ID}^T & \nu_{DD}^T \end{pmatrix} \begin{pmatrix} \bar{\phi}_x \\ \bar{\phi} \end{pmatrix}$$

or

$$\tag{27}$$

$$R_{II}\dot{\bar{q}}_{p_I} = -\nu_{II}^T\bar{\phi}_x - \nu_{DI}^T\bar{\phi}$$
$$R_{DD}\dot{\bar{q}}_{p_D} = -\nu_{ID}^T\bar{\phi}_x - \nu_{DD}^T\bar{\phi}$$

where $\bar{\phi}_x$ is the vector with elements ϕ_i^x and 0 is a matrix whose elements all vanish identically. Here R has been partitioned into

$$R = \begin{pmatrix} R_{II} & R_{ID} \\ R_{DI} & R_{DD} \end{pmatrix} \tag{28}$$

where

$$R_{II} = \begin{pmatrix} R_1 & 0 & \cdots & & 0 \\ 0 & & & & \vdots \\ \vdots & & & & 0 \\ 0 & \cdots & 0 & & R_{N_p-N_s+N_x} \end{pmatrix}$$

$$\tag{29}$$

$$R_{DD} = \begin{pmatrix} R_{N_p-N_s+N_x+1} & 0 & \cdots & 0 \\ 0 & & & \vdots \\ \vdots & & & 0 \\ 0 & & \cdots & 0 & R_{N_p} \end{pmatrix}$$

are diagonal matrices and all elements of

$$R_{ID} = R_{DI} = 0 \tag{30}$$

vanish.

4. Symmetry

We first express the rates of the arbitrarily chosen dependent processes in terms of the rates of the independent processes using the second of the stationary state conditions:

$$\dot{\bar{q}}_{p_D} = -\nu_{DD}^{-1}\nu_{DI}\dot{\bar{q}}_{p_I} \tag{31}$$

If ν_{DD} is singular we can chose a smaller set of dependent rates for which this matrix is nonsingular and express the remaining dependent rates in terms of the former. Since the following arguments do not change in this case we will assume that ν_{DD} is nonsingular.

We now use this result in the relation between the rates of the dependent processes and their conjugate forces:

$$-R_{DD}\nu_{DD}^{-1}\nu_{DI}\dot{\bar{q}}_{p_I} = -\nu_{ID}^T\bar{\phi}_x - \nu_{DD}^T\bar{\phi} \tag{32}$$

and use this result to solve for the forces corresponding to the state variables which are not associated with external sources:

$$\overline{\phi} = \left(\nu_{DD}^T\right)^{-1}\left(R_{DD}\nu_{DD}^{-1}\,\nu_{DI}\dot{\overline{q}}_{p_I} - \nu_{ID}^T\overline{\phi}_x\right) \tag{33}$$

Finally, substituting this result into the relation between the rates of the independent processes and their conjugate forces we find

$$\left[R_{II} + \nu_{DI}^T(\nu_{DD}^T)^{-1}R_{DD}\nu_{DD}^{-1}\,\nu_{DI}\right]\dot{\overline{q}}_{p_I} = \left[-\nu_{II}^T + \nu_{DI}^T(\nu_{DD}^T)^{-1}\nu_{ID}^T\right]\overline{\phi}_x \tag{34}$$

These are the phenomenological equations expressing the rates of the independent processes in terms of the forces arising from the external sources in the stationary state. They can be written in the usual Onsager form

$$\mathcal{R}\overline{J} = \overline{X} \tag{35}$$

where

$$\overline{J} = \dot{\overline{q}}_{p_I} \tag{36}$$

is the vector of phenomenological flows, and

$$\overline{X} = \left[-\nu_{II}^T + \nu_{DI}^T(\nu_{DD}^T)^{-1}\nu_{ID}^T\right]\overline{\phi}_x \tag{37}$$

is the vector of conjugate phenomenological forces. The matrix of the phenomenological resistances is given by

$$\mathcal{R} = R_{II} + \nu_{DI}^T(\nu_{DD}^T)^{-1}R_{DD}\nu_{DD}^{-1}\nu_{DI} \tag{38}$$

It is this matrix which is symmetric. Since ν_{DD} is a square matrix,

$$\left(\nu_{DD}^T\right)^{-1} = \left(\nu_{DD}^{-1}\right)^T$$

and

$$\nu_{DI}^T\left(\nu_{DD}^T\right)^{-1} = \nu_{DI}^T\left(\nu_{DD}^{-1}\right)^T = \left(\nu_{DD}^{-1}\nu_{DI}\right)^T$$

Since R_{DD} is diagonal and therefore symmetric and since B^TCB is symmetric if C is symmetric,

$$\nu_{DI}^T(\nu_{DD}^T)^{-1}R_{DD}\nu_{DD}^{-1}\nu_{DI}$$

is also symmetric. Finally since R_{II} is diagonal and therefore symmetric, $\mathcal{R}$ is a symmetric matrix.

The phenomenological forces X_i arise from the external sources. For chemical reactions they are just the overall affinities conjugate to the rates of the independent reactions, which are the rates which appear in the vector of phenomenological flows J; for electrical networks the phenomenological forces are just the net electrical potential differences conjugate to the independent currents which appear in J. For the example of Fig. 2 the chemical phenomenological forces are $\mu_a^x - \mu_{a'}^x$ and $\mu_b^x - \mu_{b'}^x$ [14, 15, 24]. Note that these forces are the net driving forces for the transport of species a and b across the membrane and that the rates of reactions 1 and 2 are the rates at which species a and b, respectively, are transported. Similarly, for the electrical example of Fig. 2 the phenomenological forces are $E_a - E_{a'}$ and $E_b - E_{b'}$, when the independent currents are taken as those through the resistances R_1 and R_2. For both the chemical and the electrical example the matrix of phenomenological resistances is given by [14, 15]

$$\begin{pmatrix} \mathcal{R}_{11} & \mathcal{R}_{12} \\ \mathcal{R}_{21} & \mathcal{R}_{22} \end{pmatrix} = \begin{pmatrix} R_1 + R_3 + R_5 & -R_5 \\ -R_5 & R_2 + R_4 + R_5 + R_6 \end{pmatrix}$$

5. Discussion

We have shown that the stationary state equations of motion for a general class of systems may be cast in the form of the phenomenological relations of nonequilibrium thermodynamics and that these relations display an algebraic symmetry arbitrarily far from equilibrium regardless of any nonlinearities present. These systems have the following properties: (1) the equations of motion can be obtained from the Lagrangian formalism extended to include dissipative force terms; (2) the dissipative forces can be expressed as products of the rates of dissipative processes and their generalized resistances; (3) the rates of change of internal state variables are related to the rates of the dissipative processes and flows through ideal sources through relations expressing conservation of extensive properties (mass, charge, length etc.). Examples of such systems include electrical networks, discrete mechanical systems and systems of chemical reactions. No restrictions are placed on the generalized resistances; they may have an arbitrary dependence on any of the variables of the system or their rates of change (i.e. flows or velocities). Thus arbitrarily nonlinear phenomena are encompassed in the general class of systems considered here. It has also been shown how the dissipative forces can be derived from a dissipation function which reduces to the classical Rayleigh dissipation function [16] when the resistances do not depend on the rates of processes but on the state of the system only.

As far as the symmetry results are concerned the work presented here can be viewed as an extension of near equilibrium Onsager reciprocity to the nonlinear, far from equilibrium regime. However, Onsager's work [2, 3] and that presented here are based on different viewpoints. Onsager's approach is microscopic and is based on studies of fluctuations around equilibrium. Subsequent attempts to extend Onsager's results beyond the near equilibrium regime were similarly based on studies of fluctuations around stable stationary states [5, 11, 12, 13]. Our viewpoint is more macroscopic and phenomenological. We assume equations of motion consistent with the Lagrangian formulation with dissipative forces of the form resistance X flow. These equations include those of dissipative Lagrangian mechanics, electrical networks and mass action kinetics of chemical reactions. Of course, these equations should find their ultimate justification at a microscopic level; however we have simply taken them to be valid on a macroscopic level here. A comparison of our results and a perturbation approach will illustrate the relation between the lack of symmetry found in fluctuation studies away from the near equilibrium regime and the algebraic symmetry found here and will show that our results reduce to Onsager's in the near equilibrium limit. We first perform a linear expansion of the general nonlinear phenomenological relations found here [(eq. (35)]. At the stationary state the phenomenological resistances $\mathcal{R}_{ij}$ can be expressed on terms of the phenomenological forces X_i and flows J_j. Expanding around a stationary state we then have

$$X_i - X_i^{ss} = \sum_j \sum_k [(\partial \mathcal{R}_{ij}/\partial X_k) J_j]^{ss} (X_k - X_k^{ss})$$
$$+ \sum_j \sum_l [(\partial \mathcal{R}_{ij}/\partial J_l) J_j]^{ss} (J_l - J_l^{ss}) + \sum_j (\mathcal{R}_{ij})^{ss} \left(J_j - J_j^{ss}\right)$$

(39)

where the superscript 'ss' indicates that the corresponding quantity is to be evaluated at the stationary state. At equilibrium this becomes

$$X_i = \sum_j \sum_k \left[\left(\frac{\partial \mathcal{R}_{ij}}{\partial X_k}\right) J_j\right]^{eq} X_k + \sum_j \sum_l \left[\left(\frac{\partial \mathcal{R}_{ij}}{\partial J_l}\right) J_j\right]^{eq} J_l + \sum_j (\mathcal{R}_{ij})^{eq} J_j$$

(40)

since all X_k^{ss} and J_l^{ss} vanish identically at equilibrium. Similarly the quantities in the square brackets are now evaluated at equilibrium and therefore vanish identically. Thus the phenomenological relations at equilibrium are given by

$$X_i = \sum_j (\mathcal{R}_{ij})^{eq} J_j$$

(41)

The $(\mathcal{R}_{ij})^{eq}$ are to be identified with the Onsager near equilibrium resistances. At stationary states far from equilibrium, however, the quantities in the square brackets do not vanish in general, and we must retain the full eq. (39). If we replace the phenomenological forces and fluxes with their deviations from the stationary state values, $\delta X_i \equiv X_i - X_i^{ss}$ and $\delta J_j \equiv J_j - J_j^{ss}$, then we can write eq. (39) in the form of the phenomenological equations

$$\delta X_i = \sum_j \mathcal{R}'_{ij} \delta J_j \tag{42}$$

but the resistances $\mathcal{R}'_{ij}$ are now constants to be evaluated at the stationary state according to eq. (39) and in general do not show reciprocity. Even if the phenomenological resistances $\mathcal{R}_{ij}$ are independent of the phenomenological forces we have

$$\mathcal{R}'_{ij} = (\mathcal{R}_{ij})^{ss} + \sum_k \left[\left(\frac{\partial \mathcal{R}_{ik}}{\partial J_j} \right) J_k \right]^{ss} \tag{43}$$

Although the phenomenological resistances $\mathcal{R}^{ss}_{ij}$ are symmetric, $\mathcal{R}^{ss}_{ij} = \mathcal{R}^{ss}_{ji}$, the resistances obtained from the linear expansion around the stationary state show reciprocity only for special cases; in general

$$\mathcal{R}'_{ij} = \frac{\partial \delta X_i}{\partial \delta J_j} \neq \mathcal{R}'_{ji} = \frac{\partial \delta X_j}{\partial \delta J_i} \tag{44}$$

The macroscopic approach taken here leads to algebraic symmetry in the phenomenological equations (35) arbitrarily far from equilibrium because these equations contain the full stationary state dynamics. In the near equilibrium limit [eq. (41)] our results are consistent with those of Onsager's microscopic approach and the symmetry becomes the differential one of Onsager reciprocity since the phenomenological equations become linear in this limit. Far from equilibrium, however, the fluctuation approach does not lead to a symmetry property since the phenomenological equations are in general inherently nonlinear in this regime, but the fluctuation approach treats only the linear approximation.

Since the approach first used by Onsager based on microscopic reversibility does not lead to symmetry far from equilibrium, one may ask what the origins of the algebraic symmetry derived here are? It arises from the form of the Lagrange equations of motion (eq. 5) with the dissipative force terms and the form of the equations of constraint (eq. 16). By writing the dissipative force terms as the products of the rates of dissipative processes occurring in the system and their resistances, we are ensured that the matrix R of the process resistances R_j is diagonal and therefore symmetric. The equations of constraint together with the Lagrangian formulation lead to the appearance in the equations of motion of the transpose ν^T of the matrix ν which appears in the equations of constraint. This property leads to the symmetry of the matrix $\mathcal{R}$ of phenomenological resistances as shown following eq. (38). (Incidentally, the symmetry of (38) is also valid for any initial matrix R which is symmetric.)

From a network point of view it can be argued that the algebraic symmetry is a topological property following from conservation laws of the form of Kirchhoff's current law (the equations of constraint) and from the equations of motion, which are of the form of Kirchhoff's voltage law. The algebraic symmetry is thus a very general property which follows from the conservation of mass, charge, etc., when the equations of motion can be given by the dissipative Lagrangian formalism. This is very different from the Onsager near equilibrium reciprocity, which follows from the principle of detailed balance at thermodynamic equilibrium when fluctuations are studied. Onsager reciprocity is more general in some fundamental sense, since it follows from a microscopic point of view. However, it could also be argued that the algebraic symmetry presented here is the more general property. It is valid for any system having the three properties enumerated above, is not restricted to the near equilibrium regime and leads to the Onsager values for the phenomenological resistances in the equilibrium limit. From this point of view the differential symmetry of Onsager reciprocity would be an equilibrium property following from the requirement of detailed balance at equilibrium.

In addition to the generality of the algebraic symmetry derived here other advantages are forthcoming. First of all, since electrical networks belong to the general class of systems treated here, any system which belongs to this class can be cast into a network form and its dynamics analyzed with standard network techniques [20]. Second, the symmetry of the matrix of phenomenological resistances decreases the amount of information necessary to determine the dynamics of the stationary state. If the matrix were not symmetric one would need to determine the functional dependencies of all n^2 phenomenological resistances for a system with n independent processes. However, the symmetry means that only $(n^2 + n)/2$ resistances have to be determined. Finally, and perhaps most importantly, the approach taken here leads naturally to a proper set of phenomenological flows and forces for which the symmetry property is valid. Eqs. (34-37) provide expressions for these quantities. As pointed out by de Groot and Mazur [4] and others, it is often difficult to find this set of flows and forces otherwise, even in the linear near equilibrium case.

Acknowledgements

The authors would like to thank S. Fassari for helpful discussions and suggestions. This work was supported by grants no. 31-29953.90 and 21-42049.94 from the Swiss National Science Foundation and grant no. 93.0106 from the Swiss Federal Office for Education and Science within the scope of the EU Human Capital and Mobilities project "Thermodynamics of Complex Systems" (no. CHRX- CT92-0007).

References

1. Callen, H. (1985) *Thermodynamics and Introduction to Thermostatistics*, Wiley, New York.
2. Onsager, L. (1931) *Phys. Rev.* **37**, 405-426.
3. Onsager, L. (1931) *Phys. Rev.* **38**, 2265-2279.
4. deGroot, S. R. and Mazur, P. (1984) *Non-Equilibrium Thermodynamics*, Dover, New York.
5. Keizer, J. (1987) *Statistical Thermodynamics of Nonequilibrium Processes*, Springer, New York.
6. Essig, A. and Caplan, S. R. (1981) *Proc. Natl. Acad. Sci. (USA)* **78**, 1647-1651.
7. Rothschild, K. J., Elias, S. A., Essig, A. and Stanley, H. E. (1980) *Biophys. J.* **30**, 209-230.
8. Stucki, J. W., Compiani, M. and Caplan, S. R. (1983) *Biophys. Chem.* **18**, 101-109.
9. Keizer, J. (1979) *Acc. Chem. Res.* **12**, 243-249.
10. Keizer, J. (1979) in H. Haken (ed.), *Pattern formation by dynamic systems and pattern recognition*, Springer, New York, pp. 266-277.
11. Prigogine, I. (1967) *Introduction to thermodynamics of irreversible processes*, Wiley, New York.
12. Glansdorff, P. and Prigogine, I. (1971) *Thermodynamic Theory of Structure, Stability and Fluctuations*, Wiley, New York.
13. Nicolis, G. and Prigogine, I. (1977) *Self- Organization in Nonequilibrium Systems*, Wiley, New York.
14. Shiner, J. S. (1987) *J. Chem. Phys.* **87**, 1089-1094.
15. Shiner, J. S. (1992) in S. Sieniutycz and. P. Salamon (eds.) *Flow, Diffusion and Rate Processes (Adv. Thermodyn. Vol. 6)*, Taylor and Francis, New York, pp. 248-282.
16. Rayleigh, J. W. S. (1945) *The Theory of Sound*, Dover, New York.
17. Goldstein, H. (1980) *Classical Mechanics*, Addison-Wesley, Reading.
18. Whittaker, E. T. (1988) *A Treatise on the Analytical Dynamics of Particles and Rigid Bodies*, University Press, Cambridge.
19. Casimir, H.B.G. (1945) *Rev. Mod. Phys.* **17**, 342-350.
20. MacFarlane, A. G. J. (1970) *Dynamical System Models*, Harrap, London.
21. Ross, J., Hunt, K.L.C. and Hunt, P.M. (1988) *J. Chem. Phys.* **88**, 2719-2729.
22. Grabert, H., Hänggi, P. and Oppenheim, I. (1983) *Physica* **117A**, 300-316.
23. Sieniutycz, S. (1987) *Chem. Eng. Sci.* **42**, 2697-2711.
24. Hill, T. L. (1977) *Free Energy Transduction in Biology*, Academic Press, New York.
25. Sieniutycz, S. and Shiner, J.S. (1992) *Open Sys. Information Dyn.* **1**, 149-182.
26. Sieniutycz, S. and Shiner, J.S. (1992) *Open Sys. Information Dyn.* **1**, 327-348.

THE PROVISION OF GLOBAL ENERGY

G.H.A. COLE
University of Hull
Hull, HU6 7RX, England

Abstract.

The global requirement for energy is increasing, partly due to an increasing world population, and traditional fuels used in the traditional way are increasingly unable to provide the energy required. It is necessary to review the sources of energy available for the future and then to develop an energy strategy to cover both local and global needs. This is the subject of this paper that offers a preliminary general assessment of the problems for providing energy in the future.

1. Introduction

It is not easy to estimate the precise figure for the current energy consumption of the world but it is likely to be about 2×10^{20} J per year [1]. With 3.15×10^7 seconds in a year, this is, in general terms, a rate of working of rather more than 0.63×10^{13} W. The world population is some 5.7×10^9 people so each person accounts for about 1.1 kW. In fact, the situation is more complicated than this. About 18% of the world's population forming the industrialised world (about 1×10^9 people) consume closely 75% of the total energy (and so using some 4.8×10^{12} W): this gives a consumption per head of population of about 4.5 kW. The industrialised world produces about 3/4 of the world's industrial output. The remainder of the world population forms the developing world and here some 4.7×10^9 people consume about 1.6×10^{12} W giving about 3.4×10^2 W each. Some statistics about the distribution of the world population and the relative areas it covers are collected in Table 1.

Here is a major problem. The industrialised world enjoys an energy usage of 13 times that of the developing world. In time this deficit of the developing world must be met by increased energy availability. For the present world population it is easy to estimate that the additional energy required will be $\approx 2 \times 10^{13}$ W. This will nearly quadruple the present energy consumption leading to an annual requirement of about 5 $\times 10^{20}$ J per annum. If this were not enough, the world population itself is growing, largely in the energy deficient developing world, at a rate of about 1.7% per annum. At this rate of increase the population will double roughly every 40 years and the world population is then set to reach 1×10^{10} people towards the middle of the next century. By the end of the century it will be greater than 2×10^{10} people. At the present rate of consumption of energy this will imply an energy usage of closely 1.5×10^{21} J per annum, or approaching ten times the present value. This will need to be greater if the

159

J. S. Shiner (ed.), Entropy and Entropy Generation, 159–173.

TABLE 1 The percentage of the world's population for the six main regions of the world together with their percentage areas.

Region	% pop	% area	Region	% pop	% area
Africa	12.1	22.3	S. America	5.6	13.1
Asia	58.7	20.3	Oceana	0.5	6.3
N. America	8.1	17.9	Europe	9.5	3.6

level of technological advance requires more energy per head of population than now. There are, however, some indications that this might not be so.

It might be thought that meeting such an increased energy demand will provide no special problems. Increasing the present methods of energy provision appropriately must surely suffice for the future. This is, unfortunately, not the case, partly because the present energy sources represent a finite energy reservoir and partly because the thermodynamic implications of present usages are crucial for the preservation of a clean environment for the increasing population. It is necessary in the longer term to change the methods of energy provision in a radical way. As a preliminary to exploring these things we must first look at the present energy sources for work transfer.

2. Present Energy Sources

There are four significant energy sources at the present time [2]: fossil fuels, biomass, hydro-power and nuclear power. The general proportions of these sources are collected in Table 2 which applies to 1989 [1]. The industrialised and developing world usages are included separately with the global mean usage.

The overwhelming importance of fossil fuels is very evident for both the industrialised and developing worlds. The use of oil and natural gas is lower for the developing world because it is often sold by these nations as a source of income. Nevertheless, fossil and nuclear fuels account for rather more than 90% of the energy sources for the industrialised world and approaching 60% for the developing world.

The substantial use of biomass by the developing world is important. This is a euphemism for brush wood and animal dung and represents a poor fuel but an important one. Its current use is a relic of the past that has no place in the modern world. The replacement will need to be a substantial source - from Table 2 it is seen that 35% of the developing world's current annual energy budget is some 1.75×10^{19} J or 5.5 $\times 10^{11}$ W.

The fossil fuel energy consumptions of three major industrialised users are listed [1] in Table 3 where data are collected for UK, USA and EU. The high use of oil and natural gas in the USA is very clear. Electricity is a most flexible source of energy and the proportions of the different fuels used to manufacture it are shown in Table 4. This Table again shows a general similarity between UK and USA in matters of energy consumption. The high proportion of nuclear sources in EU arises from the dependence of France on nuclear power (around 75% of French electricity is generated in nuclear power stations) and also the dependence of the old East Germany on nuclear sources. Regenerative sources also appear in Table 3 although only marginally.

The ways the energy is used will differ in detail from country to country but a general usage for an industrialised country is similar to the breakdown displayed in Table 5.

TABLE 2. The four main sources of energy for the industrialised and developing worlds.

Source	Percentage of whole		
	Industrialised world	Developing world	Global Mean
Fossil fuels	85%	58%	75%
coal	25%	28%	26%
oil	37%	23%	32%
nat. gas	25%	7%	17%
Biomass	3%	35%	11%
Hydro	6%	6%	6%
Nuclear	6%	1%	5%

3. The Nature of the Resources

It is interesting to review the reserve of energy expressed in Table 2 and especially to see in what way these can be expected to allow for a higher future energy usage [3]. The overwhelming fuel to meet the global energy requirement is fossil fuels although nuclear fuels can provide substantial support.. These represent a limited reserve of energy although it is not easy to determine precisely the extent of it. This is partly because finding and assessing the reserve is not always easy but also because of doubts of the economic nature of the extraction processes in some cases.

In general terms the world coal reserves (assured or suspected) are of the order of 1200 billion tonnes [1]. Taking an average coal to contain 3×10^4 MJ / tonne, this gives an energy reserve of about 3.6×10^{22} J. (The reserve for the UK is some 1.2×10^{20} J.) The world reserve for oil is about 0.96×10^{22} J while that for gas is about 0.56×10^{22} J. These are substantial reserves but are not indefinitely large when the required annual consumption of 2×10^{20} J rising to 10^{21} J is remembered. In these terms the economically important reserve of coal will last for about two hundred years but that for oil [4] and gas will last for four or five decades rather than centuries. The current trends in the consumption of the fossil fuels will require changes in the energy source for work transfer over the next 30 to 40 years. Transport is a particularly sensitive area in this connection. The need for an alternative fuel to petroleum is clear.

Nuclear fuels offer other time scales [5]. Used in a thermal reactor, 1 kg of uranium has the energy value of 1.2×10^3 MJ (equivalent to 20 tonne coal), while used in the fast reactor 1 kg uranium has the energy value of 6×10^7 MJ (2000 tonnes coal equivalent). These are very substantial reserves and would be capable of meeting the global energy requirement for many centuries.

The developing world uses a high proportion of biomass but this has a low specific energy content [3]. It was seen in Section 2 that the total available energy is about 1.75×10^{19} J. It will be seen in Section 6.2 that this is about the energy provided by photosynthesis covering some 10% of the world land surface and can hardly be increased further - indeed, the brushwood areas of the world are tending to decrease with deforestation rather than increase. The availability of biomass has probably reached a saturation value.

Hydro-power involving substantial dam structures represents a difficult and financially costly approach to energy provision. The essential difficulties are associated with geological structures to contain the water and ecological damage that might follow its collection.

TABLE 3. The Energy Consumptions of United Kingdom, European
Union and United States of America for 1989 (OECD Statistics)

Region	Energy Consumption (units - 10^{18} J			
	Coal	Oil	Nat. Gas	Electricity
UK	0.5	2.8	2.0	1.0
EU	2.2	17.5	8.1	5.3
USA	2.4	29.0	15.1	9.2
Total	5.1	49.3	25.2	15.5

4. Thermodynamic Consequences

The 1st and 2nd laws of thermodynamics place severe constraints on the conversion of heat energy to work energy [6]. The second law requires any engine working between finite temperatures to reject heat energy - it is not possible to convert all of a given quantity heat into work. The thermal efficiency of the engine (= work output / heat input) must be less than unity. The central question then for a practical engine is how much is the thermal efficiency below the ideal set by the Carnot engine working between the same temperatures as the real engine. In practice, the thermal efficiency of the transfer of energy from one form to another is low, generally less than 50% and often much lower. Most of an initial energy reservoir is rejected as heat to the environment and the work output is correspondingly low. Environmental warming is a central part of the provision of work.

The effectiveness of the work output is more difficult to assess. Whereas the Carnot engine works with heat reception and heat rejection at fixed upper and lower temperatures, such isothermal heat transfer is not available in the real world for real engines. Heat must be "driven" through a temperature gradient and the usual conditions are those of constant pressure [7]. This is especially the case for the Rankine and Joule thermodynamic engines using respectively steam and gas as the working fluid.

The means of providing the upper temperature and perhaps sustaining the lower one is the fuel. Energy is extracted through some combustion process and the incombustible remains forms the waste. The engine as an entity rejects both heat energy and waste matter (which is generally called pollution, especially if it is not to our liking - gold is a pollution from supernova explosions but we don't generally refer to it as a pollutant). The engine rejects both heat and waste materials and we might say quite generally

Every continuously working engine must have a hot active exhaust system

In energy terms, any method of converting heat energy into work energy must heat the environment to some extent and produce some pollution. It may be that these are near where the work is to be performed or they may be a lesser or greater distance away, but they must be there somewhere. In general, the high density fuels (fossil fuels and nuclear) have the products of conversion essentially where the work is performed. The other sources, low density as we shall see later and called regenerative sources, have the heat rejection and pollution elsewhere separated widely in both space and time from where the work is released. This is an enormous advantage if these sources are strong enough for our purposes.

TABLE 4. The approximate proportions of different fuels used to manufacture electricity in three industrialised regions.

Region	Energy Source			
	Fossil fuels	Nuclear fuels	Hydro	Regenerative
UK	75 %	23 %	2 %	--
EU	56 %	≈ 36 %	8 %	≈ 0.2 %
USA	72 %	19 %	≈ 9 %	≈ 0.3 %

5. Four Basic Sources of Energy

There are four sources of energy that are basic to practical energy provision [3]. These have their origins in

♦ The Solar Radiation
♦ Energy in the Ocean Tides
♦ The Formation of the Earth
♦ Nuclear Rearrangements and Radioactivity

Each is a cosmic source whose strength is outside our control. It is interesting that each is associated with the action of gravity either directly or indirectly. Solar radiation (Sections 6 and 9) arises from the gravitational compression of the solar materials; the ocean tides (Section 8) are the direct consequence of the gravitational interaction between the Earth, Moon and Sun; the Earth formed (Section 7) by collapse due to gravitational attraction between the primordial particles and gases; and the heavy elements are formed in supernova explosions due to a gravitational instability in heavy stars as the stellar fuel is used up.

It is interesting to ascertain the energy involved in each source and to estimate in what way it could contribute to the global energy requirement.

6. The Sun as a Source

The Sun emits 3.90×10^{26} W continuously (the Solar Luminosity) corresponding to a black body at a temperature of 5800K [8]. This energy is derived from the conversion of hydrogen atoms into helium atoms in the central region of the Sun (see Section 9). Helium is the pollution in the process and this stays at the solar centre. The Earth orbits the Sun in a closely circular orbit with mean radius 148 million kilometres. The solar radiation flux in the vicinity of the Earth (the Solar Constant - even though it is not strictly constant) is then about 1.4 kW / m^2. Not all of this flux is absorbed by the Earth. The Earth's albedo is 0.3 so about 30% of the radiation is reflected immediately into space: this leaves 980 W / m^2 to enter the Earth's environment, which is closely 1 kW / m^2. This quantity is divided unequally between the surface (land and sea) and the atmosphere. About 80% (800 W / m^2) is absorbed by the surface leaving 20% (200 W / m^2) to enter the atmosphere. The solar radiation heats the surface to a mean temperature of about 290K [9] and this is the temperature of the Earth measured from space. It is interesting to notice that the mean temperature of the Earth in equilibrium with the Sun would be about 265K if it had not got an atmosphere or a little below the freezing point of water. The effect of the natural atmosphere is to give a higher temperature where life can begin and develop on Earth.

TABLE 5. A typical distribution of the use of
energy by an industrialised country

User	Percentage %
Domestic / Commercial	25
Transport	25
Industrial	30
Transmission / Distribution	18

6.1 DIRECT COLLECTION OF RADIATION

The continental surface of the Earth is 1.48×10^{14} m^2 and the radiation falling on that area is 1.2×10^{17} W. The estimate was made of an energy usage of 10^{14} W during the next century was made in Section 1 and it would seem in principle that this could be obtained by collecting radiation over an area of about 0.1% of the continental surface or about 1.5×10^{11} m^2 [3, 10]. This is similar to the land area of the UK but, of course, collecting would need to be done in the equatorial regions. This expectation of principle is, unfortunately, restricted in practice for a number of reasons. These are
♦ The efficiency of the conversion process $\beta(c)$ to produce energy in some other form;
♦ The Sun doesn't shine for the full 24 hours of the day - on average solar radiation is collected for 0.5 day;
♦ To cater for the period when the Sun is not shining it is necessary to have a storage system that can be used when it is dark. This is very serious because the main land masses of the Earth are all contained within one hemisphere. We might suppose one half the collected radiation goes to storage;
♦ The collectors, whatever they may be, must be serviced and replaced giving a down time d as a fraction of the day;
♦ The energy obtained this way must be distributed (most likely as electricity) and there is a loss $\beta(d)$ associated with that.
♦ There will also be a factor b representing the fraction of the possible total radiation that is actually obtained due to misalignment of the collectors.
Consequently, the energy flux E actually obtained from the collection of solar radiation over the fraction A of the surface area can be calculated from the expression

$$E = 1.2 \times 10^{17} \ 0.25 \ \beta(c) \ [1 - d] \ [1 - \beta(d)] \ b \ A \quad (W)$$

If we take $d = 0.33$, $\beta(d) = 0.15$ and $b = 0.9$ we find for the energy flux

$$E = 1.5 \times 10^{16} \ \beta(c) \ A \quad (W) \tag{1}$$

This is, in effect, the maximum energy flux that can be expected to be obtained by any particular array for any given area A.

6.1.1 Photovoltaic Conversion
The solar radiation falls on a flat semi-conductor surface arranged normal to the direction of the rays [11]. The material then converts the solar radiation into electricity. There is a theoretical maximum conversion efficiency (about 27%) determined by the quantum band gap structure of the semi-conductor lattice. The material is very costly to make and handle and at the present time the most commercially attractive materials have a conversion efficiency of about 7%: for our

present purposes of estimation let us call this a 10% efficiency. Setting $\beta(c) = 0.1$ we find for the energy flux

$$E = 1.7 \times 10^{15} \, A \quad (W)$$

so the "target" of 10^{14} W could be met with $A = 0.1$ which means with collectors covering 10% of the Earth's surface. The present requirement, could it be met this way, would require some 4 or 5% of the land coverage. From Table 1 we see that is would be a land area the size of Europe - for the longer term it would require a land area half that of Africa. A storage facility would, of course, be needed. The associated engineering tasks if this is to be brought about must be quite considerable.

6.1.2 Direct Collection by Radiation
Another possibility is the focusing of solar radiation by a mirror system to heat a fluid which, through a heat exchanger, will convert water into steam [3]. The work output is then obtained using a Rankine cycle in the conventional way. The steam temperature may not be high by this method and the thermal efficiency will be only marginally greater than that of photoconversion (Section 6.1). There is still the storage requirement to be met. It could be possible, of course, to mix firings to have the solar during the day and a fossil / nuclear source during the hours of darkness. This introduces the idea of a mixed fuel firing to avoid storage problems. The power plant would need. of course, to be in the sunny equatorial regions.

6.1.3 Passive Collection
We might mention here the passive heating of the space enclosed by a volume as in a greenhouse [12]. It does not produce energy for work but it does reduce the energy requirement. According to Table 5 about 25% of the energy used in an industrialised country is involved with general domestic and commercial purposes. Careful architectural design of buildings can reduce this energy requirement by as much as 90% thus leading to a substantial energy saving: this could be as much as $((0.9 \times 2) / 4) \times 10^{20} = 4.5 \times 10^{19}$ J per year. This could go a very long way towards supporting the increasing energy requirement for the developing world. Unfortunately, the task of converting the existing houses is not possible at a stroke!

6.2 PHOTOSYNTHESIS

Living materials are the most characteristic feature of the Earth within the Solar System and the flora live by photosynthesis [13]. This converts solar radiation into an energy form that can be released by burning the flora material. The particular form of interest is trees which contain large amounts of this energy.

6.2.1 Modern Radiation
The same general analysis of Section 6.1.1 will apply here [3], the quantity of energy again being controlled especially by the efficiency of the photosynthetic conversion process. This is, in fact, extraordinarily small - no greater than 1% in general. The eqn (1) will apply again except that $\beta = 0.01$ and a factor must be introduced to account for the time it takes to grow the flora to be converted to energy. Trees give the greatest possibilities but these take several decades to grow - let us suppose 10g years. For a

steady supply, with replanting balancing burning, it is possible to burn 1 / 10g of the area of trees annually so the maximum rate of energy production obtainable follows from (1) as

$$E = 1.5 \times 10^{16} \times 0.01 \times A / 10 \text{ g.} \quad \text{(W)}$$

Taking g = 5 (a not unreasonable value for a tree of good quality but the value is not critical) we find

$$E = 3 \times 10^{12} \quad \text{(W)} \tag{2}$$

with A = 1, that is all the surface covered. The estimate in Section 2 of 5.5 x 10^{11} W from this source is obtained from about 20% of floral land coverage. The energy available here is limited and the present use of biomass by the developing world probably represents about the maximum that can be obtained from this method.

6.2.2 Ancient Radiation

Photosynthesis occurred in the past in unique geological conditions that existed several hundred million years ago, and the conditions themselves lasted for two hundred million years. Trees growing then in warm swamp conditions (fern like and rather unlike trees growing today although there is one Australasian fern that is related) fell as they died and underwent a metamorphosis with pressure and temperature to form coal. The expression (2) applies for each of some 2 x 10^6 years when the seams were laid down to provide a "frozen" energy source that will release its energy through combustion. The maximum potential source, probably covering about 10^{-4} of the Earth's surface, is some 3 x 10^{12} x 2 x 10^6 x 10^{-4} = 6 x 10^{14} W. The extreme compression means that the energy is now a high density source. A comparable metamorphosis occurred in the seas where the decay of elementary marine creatures formed oil and methane (natural gas) again as concentrated energy sources of broadly the same energy capacity as coal.

6.3 WIND POWER

Everyone knows the great power that can be shown by the wind and this has been used throughout recent history as a means of power for driving ships at sea and machinery on land. The task of harnessing this on an industrial scale is beginning to be undertaken in earnest.

The wind enters a "machine" [14], the linear wind energy being translated into some drive energy generally involving rotation. The modern applications envisage the production of electricity through a dynamo.

The machine will generally have the form of a cylinder with a blade (propeller) to interact with the wind. It will be realised that not all the energy can be removed from the wind: if this were done the air would be stationary and the machine could work no more! Nevertheless, to gain ideas of orders of magnitude we will suppose that all the energy in the air is available for extraction.

If the wind speed is v m/s and the air density is ρ the energy in unit volume of air is $E = (\rho/2)v^2$. The quantity of energy flowing per second through a unit area perpendicular to the wind direction is then E v = $(\rho/2)v^3$. The energy that can be extracted per unit time is proportional to the cube power of the wind speed. The device has been presumed cylindrical and we suppose the radius is L giving a cross-sectional area πL^2. The energy entering the device per unit time is then

$$E(w) = (\rho/2) \, v^3 \, \pi L^2.$$

This is for one device. If there are n(w) devices spread per unit area and distributed over an area A the energy flux that can be obtained E(o) is

$$E(o) = \pi(\rho/2) \, v^3 \, L^2 \, n(w) \, A.$$

This energy must be converted into that useful for work and so must be subject to the action of several practical factors. These are

◆ the efficiency of the conversion process $\beta(w)$
◆ the down time d
◆ the transmission loss factor $\beta(d)$

so that the energy E actually obtainable is

$$E = E(o) \, \beta(w) \, [1 - d] \, [1 - \beta(d)].$$

The number of devices that can be placed per unit area is dependent on the size of the orifice L. Each device will cause a turbulent wake to form behind it and this is not suitable as the initial flow for another device. Again, the extent of the wake is greater the larger the orifice size (that is the larger L). The larger the value of L the greater the engineering problem involved in constructing the device and the greater the financial cost. Clearly there must be a compromise. There is no simple relationship between L and n(w) but for L = 20 m it seems that $n(w) = 2 \times 10^{-4}$, which is about 4 machines per hectare.

If we suppose, as typical values, $\rho = 1$ kg / m^3, v = 5 m/s, $\beta(w) = 0.9$, $1 - \beta(d) = 0.85$, L = 20 m, 1 - d = 0.67 and $n(w) = 2 \times 10^{-4}$ we find E = 8 A. For a target figure for 2050AD of 10^{14} W we must have $A = 10^{14} / 8 = 1.2 \times 10^{13}$ m^2 which is very nearly 10% of the Earth's surface. The present requirement of a few times 10^{13} W could be met by spreading the devices over a few percent of the Earth's surface but, of course, the machinery is not yet available. There will be the need to have a strong storage system to cater for periods when the wind is not strong.

6.4 WAVE POWER

The power of the sea is legendary but so also are its periods of calm. Although the energy obtained must be accepted as being variable we can associate a mean value for the energy e of a wave profile per unit length [3]. If appropriate devices are deployed [15] to extract the energy from the waves energy can be obtained. If the length of the deployment is L, the maximum energy E obtainable is E = e x L. This is subject to all the usual losses in practice but it must also be admitted that the engineering at sea would require especial skills that are not yet available. To have some idea of the energy involved, we might take e = 20 kW / m as a broadly typical energy flux per unit length of wave. For a length of collector 1.6×10^6 m (which is roughly the length of the English/Scottish North Sea coast, then $e = 3.2 \times 10^{10}$ W. This value would need to be reduced in practice by downtime (particularly important for marine engineering), conversion losses (so far suggestions here have been grossly inefficient) and transmission losses (the electrical engineering associated with these devices would be very difficult). There has been little progress so far on any practical devices for energy conversion. The engineering must be of the heaviest to withstand periods of storm which is not efficient under more normal conditions. A deployment of collectors would also stop shipping crossing the system. The obtainable energy would hardly satisfy the global energy need. While it is useful to continue to study these problems there would seem little chance of this method of energy conversion ever providing a viable energy source on its own.

6.5 THE HYDRAULIC RESOURCE

There are lakes and regions of water throughout the land mass of the Earth, the water coming from precipitation (rain) and draining to the sea by way of rivers [3]. The water is at various heights above sea level and the move to the sea represents a fall through the gravitational field of the Earth. The energy appears as the rapid motion of the water and it would be possible to convert the energy to work using a dynamo system.

The world hydraulic resource is quoted by United Nations estimates as 1.03×10^{20} J. If this were drained into the oceans in 1 year the rate of working would be $\approx 3.3 \times 10^{12}$ W. Replacing the reserve will provide a continuous source of energy but it is likely to take more than one year to replenish the lakes and reservoirs. Hydrologists suggest it may take as much as 5 years to do this which would give the rate of working of $\approx 7 \times 10^{11}$ W. This is not a significant fraction of the global requirement now leave alone for the future. We can conclude that the hydrological reserve may have local importance in a few places but has no global significance. The initial world estimate of the reserve would need to be too low by several orders of magnitude if this conclusion is to be challenged.

6.6 OCEAN THERMAL POWER

The surface waters of the oceans are at a higher temperature than the water deeper down. There is the possibility here of driving an engine by the difference of temperature. The difference is in practice small - perhaps 20K at the very most - but this difference is spread over perhaps 1 km, giving a temperature gradient of 2×10^{-1} k/m. There are no global possibilities here although there could be some very local interest for very low power projects.

7. Geothermal Energy

There is a surface temperature gradient of about 30K / km in the near surface regions of the Earth and the higher temperatures below the surface lead to geysers and similar phenomena at certain places (in some inter-plate regions). The heat distribution is far from uniform and in some instances the temperature gradient is larger than at other places. If water naturally accumulates there deep down it can be turned to steam and the resulting local pressure will force the steam to break the surface and form a geyser. These natural water spouts are known in several places on the Earth, all associated with boundaries between contiguous geological surface plates or geologically active regions (such as in California and in Japan). Some of these have been harnessed to provide superheated steam which provides typically some 500 kW plant using the Rankine cycle. There might be perhaps a dozen such plant world wide and, although they will make a local contribution to the local energy budget, they are not of global significance.

It has been suggested that a fluid (for instance water) could be piped down to these regions, heated up, and used hot at the surface, like an artificial geyser. It is necessary to pipe water down to considerable depths but the returns in work transfer and in financial economy would not be very impressive.

8. Tidal Power

The twice daily tides in the seas and oceans offer the opportunity to gain energy from the gravitational attraction of the Moon as a primary source and the Sun as a secondary one [16]. The rising waters are allowed to enter and enclosed volume through a turbine to produce electricity and then the water is allowed to run out again when waters have fallen for the second half of the tide.

The tidal range is far from uniform over the Earth, the sea floor and land topologies being a most dominant factor here. The North Sea / English Channel coasts are particularly favoured and a tidal range of 10 m can be encountered. The energy that can be expected to be obtained is, unfortunately, not impressive. For example, if all the possible sites around the UK coastline were used the total energy provided has been estimated as about 2×10^{10} W and for this all the estuaries would have to be either closed or heavily restricted to shipping. The French experience has been similar. The cost per unit of electricity is likely to be uneconomic for the foreseeable future.

9. Fusion Power

The solar radiation comes from the ultimate energy source which fuels the stars themselves [13]. In its simplest form, found in the lower mass stars like the Sun, this involves the conversion of hydrogen into helium - four hydrogen nuclei (protons) join to form one helium nucleus. In general, when the density of hydrogen atoms is high enough and their energy is large enough, the atoms become ionised and four protons (hydrogen nuclei) can combine to form one helium nucleus. The helium nucleus contains less energy than the four constituent protons and the excess energy is released. The energy released is about 4×10^{12} J per helium atom formed or about 1×10^{12} J per hydrogen atom destroyed: expressed alternatively, this is about 6×10^{14} J / kilogram.

These processes can occur and be self sustaining only if the density of the hydrogen gas is high enough and if the temperature of the hydrogen atoms (and so their kinetic energy of motion) is large enough. This occurs naturally at the centre of a star if the temperature exceeds about 16×10^6 K and the pressure exceeds about 40×10^6 atmospheres. This is true for aggregates of hydrogen with masses in excess of about 0.6 of a solar mass.

The reaction is simplest for the smaller masses where protons are converted directly into helium following the so-called p-p reaction. If four protons (4 x p) are to be converted into two protons and two neutrons (the helium nucleus) the two neutrons (2 x n) must be formed from two protons according to the scheme

$$p = n + e^+ + \gamma + \nu$$

where γ is gamma radiation and ν represents neutrinos. This relation involves the so-called weak nuclear interaction which is relatively slow in time. The full scheme is

$$^1H + ^1H = ^2D + e^+ + \nu$$
$$^2D + ^1H = ^3He + \gamma$$
$$^3He + ^3He = ^4He + ^1H + ^1H.$$

The whole conversion process will require about 0.75×10^6 years. There is a slightly different process for higher mass stars in which the hydrogen is converted to helium but using lithium and beryllium as catalysts. This applies at a rather higher temperature - the nuclear reactions are very sensitive to temperature.

The energy released at the centre of the star will be partly in the form of the kinetic energy of the particles that result from the interaction and partly in electromagnetic radiation. This will be predominantly in the X-ray region at the temperature of the centre. In moving to the surface the photons suffer collisions and lose energy at each one (this is the Compton scattering). Moving through the main body of the star in a Brownian motion they emerge at the surface with a mean photon energy in the optical region. Some photons will have traversed the distance more quickly than the average and will have a higher energy while others will have had more collisions than the average and so will have less energy - this provides the Planck black- body radiation curve.

There has been the desire to establish such an energy source in a controlled manner on Earth [17] since its validity for stars was finally established in the early 1940s. It will be realised that the laboratory / engineering rig version will need to be very different from that applying to stars directly. To begin with the time scale must be shorter - energy is required now and not in a million year's time. Again, it is not desirable that the main radiation produced should be in the form of X-rays. Achieving the appropriate density and temperature proves to be very difficult - in fact it has not yet been achieved in a controlled way.

The long temperature scale can be avoided by cutting out the step involving deuterium and the weak interaction: the starting material can be a mixture of hydrogen and a deuterium or lithium salt. This will mean the reaction temperature will need to be increased to perhaps 100 million K. There are two different conditions that can be considered - the controlled and the uncontrolled production of the energy.

For uncontrolled conditions the high temperature can be obtained from the explosion of an atomic bomb, where the central temperature is comparable to that at the centre of the sun. There are problems here of containing the vaporised material together long enough to allow the reaction to occur and, to achieve maximum energy output, the triggering of the volume simultaneously. These are the considerations involved in the thermonuclear hydrogen bomb.

Controlled conditions with the production of energy have not yet been achieved and the problems involved can be guessed from what has been said so far. If this is the energy source of the future it will take a long time to reach the research and development stage and longer again to overcome the many engineering problems that would be involved in its commercial exploitation. One major problem would be the production of high energy neutrons. These will interact with the material comprising the "fusion reactor" making it radioactive - it will also weaken the structure. There must be a long time scale before fusion machines can be developed as safe, commercial entities.

10 Comments and Summary

The provision of work from an initial energy source is seen to involve a certain heating of the environment and the production of a certain amount of pollution, some of which may be harmful. It is important, therefore, the assess which energy source provides the minimum of these effects and which is the most possible for use in practice. We see that, as is usual in the real world, the answer is really a compromise.

There is an initial choice between two alternatives. One choice is to make an energy requirement that corresponds to the way we would wish to live. This involves at the present time a few times 10^{13} W as a global requirement. This is likely to increase to some 10^{14} W over the space of the next 50 years. The second choice is to accept an energy provision associated with a particular technology or range of technologies and rearrange global living to fit that energy level. This has not, so far, been suggested as an energy / living strategy for the future.

The analysis of the preceding Sections will allow us to cover both approaches. It is important to realise that our analysis gives estimates of the maximum energy that can be expected from each source.

In making these analyses it is important to distinguish between high density sources of energy (where the source can be very compact) and low density sources (where extended areas are important). In practice this separates fossil fuels and nuclear sources on the one side from the remainder on the other. The high density sources have the associated pollution at the surface of the Earth whereas the low density sources, called regenerative for obvious reasons, have the associated pollution elsewhere, for instance in the centre of the Sun. It is a remarkable effect of gravitation that it shows no pollution at all.

10.1 REGENERATIVE SOURCES

In very general terms, the major regenerative sources, that is wind and solar, require a substantial minor fraction of the Earth's surface to provide the required energy. For the future this proportion of the surface may need to be 10% or even more. The land surface will also have to be in the appropriate geographical location, in the tropics for the solar reception or in the windy places for wind power. The transmission problems will be huge quite separately from the political and social problems involved with committing so much land to a single supra-national purpose. Both wind energy and solar energy sources require the existence of a substantial high power storage facility which does not yet exist. They can, however, meet the global power requirement now and in the future if sufficient land is available for the display of solar cells or wind collectors. They may not, however, have sufficient potential to provide the power for local regions if this requirement is large. As an example, they could not provide more than about 20% of the power requirement of the UK and perhaps less for some other northern countries. Nevertheless, in global terms, these energy sources must be regarded as long term possibilities. Wave power, biomass, geothermal and tidal sources have been found all to fail to provide sufficient power for present or future purposes. All the regenerative sources, therefore, fail to meet the energy requirement even though they may have great attractions form the point of view of minimum material pollution.

10.2 HIGH DENSITY SOURCES

The high density fuels provide the major contribution to the global power now and could meet the full requirement both now and in the future. They are, however, associated with problems of pollution, which may be severe, and the future availability of the fuel can be open to question. To explore these aspects it is necessary to separate fossil from nuclear fuels.

The fossil fuels, being composed of hydrocarbon molecules, lead to a pollution comprising particularly CO_2, H_2O and CO. The impurities in the fuel include sulphur which leads to SO_2 some of which combines with water to form sulphuric acid. This has some desirable features within a coal fired plant but is harmful outside when it falls as a fine rain. It is essential to withdraw the impurities from the combustion products although it is not easy to do this. If the elementary products do not go to the environment they must be collected in some way and large quantities are involved: all the chemical elements present in the unburned fuel are to be collected (in a different chemical arrangement) after combustion. Having been collected they must be disposed of harmlessly - CO_2 is one example of a material that is extremely difficult to dispose of harmlessly. Special forms of combustion, such as pressurised fluid bed combustion, may not allow impurities to get to the atmosphere but the collected impurities need to be disposed of. This is also true of gasification processes where coal is converted to an artificial gas which is then burned. The disposal of waste is a major problem for fossil fuel working.

Nuclear fuels give off no gases to pollute the atmosphere and so to heat the environment. The fuel after "combustion" must be disposed of specially because it is radioactive. Some of the radioactivity is strong and will last for hundreds or thousands of years. Others are weak and short lived. The most effective present method of disposal is burial and there is no scientific reason why this should not be safe and acceptable. The technology is based on the isotopes and derivatives of uranium which is also an element which has been the basis of military technology of the most horrendous form. Provided the technology is treated professionally and carefully it has a harmless application to civil power provision.

The waste products (which have a role in achieving stability in the reactor and which would have enormous uses if they were not radioactive) would be best if they could be transmuted to other elements which are either very weakly radioactive or not radioactive at all. There is considerable research being conducted in several countries which aims to achieve this but the priority seems not to be high. It is highly desirable to develop a nuclear technology not based on uranium directly and a thorium technology is possible for the future. It would centre on forming the isotope ^{233}U from thorium, holding it in sub-critical amounts until power is required when the quantity of ^{233}U is increased by accelerator methods to achieve criticality. This is still a proposal to be investigated but does point the way towards a better nuclear technology. The quantity of thorium in the near surface of the Earth is four times that of uranium and would provide a source of power able to satisfy global needs for many centuries.

10.3 CONCLUSION

So where are we now? It is clear that the present methods must form the basis of power provision for the immediate future. Techniques must be introduced to minimise the atmospheric pollution. While this is the case it is essential to save energy as much as possible and to make decisions on the future energy requirement to be provided. The arguments of the previous sections show that regenerative sources are not realistically sufficient to provide the energy that is likely to be required. Fossil fuels have limited life times. Nuclear sources could be capable of providing all the energy that we can require but need modifications to the technology.

So, do we set our energy requirement irrespective of methods of provision or do we use the energy available from a limited range of socially acceptable sources? It is important that this matter be decided soon. Either choice can involve sources that are safe if professionally controlled and run. This is not a stable situation and the matter requires the most urgent attention. There seems a remarkable lack of recognition world wide that there is here an essentially crisis situation.

References

1. The data quoted here come from various publications of the United Nations: *e.g. World Statistics in Brief*, (1992) United Nations Pocketbook (14th Ed); Reports from the OECD secretariat and from the annular economic review published by *The Economist* (London). For Europe see *The Guiness European Data Book*, (1994) Guiness Publications Limited, London. The energy statistics for the different European countries are presented in a series of books by Europa Publications, London.
2. McMullan, J.T. (1983) *Energy Resources*, Arnold, London (2nd Ed); Adelman, M.A. (1983) *Energy resources in an uncertain future*, Ballinger, Cambridge, Mass.
3. Cole, G.H.A (1992) *Energy World* No.199, 15 - 19, Institute of Energy, London; (1993) *Physics Review* 2, 12 - 18, Philip Allen, Oxford.
4. *Oil Map of the World*, (1958) Daily Mail, London.
5. see e.g. Bennet, D.L.(1981) *The elements of nuclear power*, Longman, London (2nd Ed).
6. Rogers, G. and Mayhew, Y. R.(1992) *Engineering Thermodynamics*, Longman, London.
7. Cole, G.H.A. (1991) *Thermal Power Cycles*, Edward Arnold, London.
8. Mitton, S. (Ed), (1977) *The Cambridge Encyclopaedia of Astronomy*, Cambridge University Press, Cambridge, Ch 8.
9. Houghton, J.T. (1979) *The physics of atmospheres*, Cambridge University Press, Cambridge.
10. Cheremisinoff, P.N. and Regino, T.R. (1978) *Principles and applications of solar energy*, Ann Arbor Science Publications, Ann Arbor.
11. Fahrenbruch, A.L. (1983) *Fundamentals of Solar Cells: photovoltaic energy conversion*, Academic press, New York.
12. Howell, D. (1979) *Your solar energy home*, Pergamon Press, Oxford.
13. Shu, F. H. (1982) *The Physical Universe*, Oxford University Press, Oxford.
14. Eldridge, F.R. (1980) *Wind Machines*, Van Nostrad Reinhold, New York (2nd Ed).
15. McCormick, M.E (1981) *Ocean Wave Energy Conversion*, Wiley, New York.
16. see e.g. French, A.P. (1971) *Newtonian Mechanics*, Norton and Company Inc., New York.
17. Dean, S.O. (1981), *Prospects of Fusion Power*, Pergamon Press, New York; Dolan, T.J. (1982), *Fusion Research - Principles, experiments and technology*, Pergamon Press, New York; Hunt, S.E. (1980) *Fission, fusion and the energy crisis*, Pergamon Press, Oxford.

MULTIPLE SOURCE PHOTOVOLTAICS

This paper is dedicated to the memory of Professor John Parrott who died on 25th November 1995

P.T. LANDSBERG

Faculty of Mathematical Studies, University of Southampton, SO17 1BJ, UK

A. DE VOS

Vakgroep voor Elektronika en Informatie Systemen (ELIS), Universiteit Gent, B-9000 Gent, Belgium

P. BARUCH

Groupe de Physique des Solides (associé au CNRS), Université Paris 7, F 75251 Paris Cedex 05, France

AND

J.E. PARROTT

Department of Physics and Astronomy, University of Wales College of Cardiff, P.O. Box 913, Cardiff CF2 3YB, UK

Abstract.

The work flux output of a solar cell subject to n black-body pumping sources is optimised in an approximate sense (investigated in earlier papers). Some theoretical efficiency curves are shown for a three-source situation. The various resulting efficiencies are analysed and compared. Further, an appropriately generalised solar cell equation is found and discussed. A relation with the generalised Shockley-Queisser efficiency is also established for the new extended setting. The one, two and three dimensional cases are covered in a single formalism.

1. Introduction

A normal solar cell is illuminated not only by the sun but also by diffuse radiation from the sky, from partial cloud cover and from various reflections. Unfortunately there are so many disturbing interactions involved in the radiation balance of the earth that "no simple theory can be expected

175

J. S. Shiner (ed.), Entropy and Entropy Generation, 175–195.
© *1996 Kluwer Academic Publishers.*

to give an accurate quantitative statement" [1]. These additional sources
of radiation are here modelled (to achieve merely a first orientation) by
black-body sources such as the sun, a clear sky and a few clouds. Consider
the "approximate" maximum of the solar cell output power (or efficiency
η) with respect to the energy gap E_g of the semiconductor for a given
illumination and for given other external factors. The word "approximate"
is here used in the sense that one uses $\partial \eta / \partial E_g = 0$ for fixed voltage V,
without proceeding to the condition $(\partial \eta / \partial V)_{E_g} = 0$ which is (strictly) also
required. This procedure is justified in fig.3 of ref. [2]. For a solar cell whose
temperature is equal to that of the surroundings $(T_c = T_s)$ and which
is hemispherically surrounded by a single pump at temperature T_p (this
corresponds to the so-called 2π-geometry) one finds for this approximate
case a result given in Section 3 of [3]:

$$qV = E_{g,app} \left(1 - T_s/T_p\right) \tag{1}$$

Here $E_{g,app}$ is the energy gap required for this approximate maximum. This
observation is of course very suggestive because of the appearance of the
Carnot factor involving the temperatures T_s and T_p of the surroundings
and the pump respectively. It arises from rather general considerations.
For example, it is independent of the density of states [3]. The question
arises what happens if one has several (n, say) black-body pumping sources
of "strengths" Γ_i and absolute temperatures T_i ($i = 1, 2, ..., n$). This paper
seeks to investigate this matter.

The conclusion is that the Carnot-type formula (1) cannot be maintained for $n \geq 2$. See equations (10), (11) and (20) below, for the replacement formulae. In some cases it does not even hold for $n = 1$; see equation
(18), below.

Of course the approximately maximal work flux output for the generalised model (see Section 2) leads one naturally to ask about the resulting
efficiency (see Sections 4,5), including the approximately optimal efficiency.
For the latter a simple inequality is derived in equation (48).

Some general remarks about the solar cell equation are added at the
end of the paper (Section 7), and entropy generation in a cell is discussed
in Section 8.

Solar cells are of course always contemplated in $D = 3$ dimensions, so
that the generalisation to D dimensions in Appendix I and in Sections 5-7
may be regarded as perverse. However this generalisation is "trivial" and
a reader can easily put $D = 3$ where needed. Also the useful cases $D = 1$
[4] and $D = 2$ can readily be obtained from the general formulation, but
would not be available, if only the case $D = 3$ were discussed.

2. Approximately Optimum Work Flux Output

The number flux of electron-hole pairs is due to the number flux of photons of energy E exceeding the energy gap E_g. If l is 1 or 2, depending on the number of directions of polarisation, the photon number flux per unit solid angle in range dE is

$$\tilde{g}_3(E)\,dE \equiv \frac{lE^2dE}{h^3c^2} = \frac{l(kT)^3}{h^3c^2}x^2dx \equiv g_3(x)\,dx \equiv A_3(kT)^3x^2dx \qquad (2)$$

where $x \equiv E/kT$ and T is a normalising temperature. In D dimensions ([5], p.1077), with $\Gamma(..)$ denoting the gamma function,

$$g_D(x)\,dx = \frac{l\pi^{(D-1)/2}(kT)^Dx^{(D-1)}dx}{\pi\Gamma\left(\frac{D+1}{2}\right)c^{(D-1)}h^D} \equiv A_D(kT)^Dx^{(D-1)}dx \qquad (3)$$

Consider a black-body source at temperature T_p (the "pump") with a geometrical factor Γ_p given for $D = 3$ by $\int\int\cos\theta d\omega/\pi$; it is the fraction of the solid angle subtended by the pump and must be distinguished from $\Gamma()$ (with brackets following it) which is the gamma function. For sources covering a small part of the sky, this is approximately proportional to the solid angle Ω subtended by the source, multiplied by the cosine of the angle of incidence of the light on the solar cell $\pi\Gamma \sim \cos\theta\int\int d\omega \sim \Omega\cos\theta$ (see Appendix I). The theory of the Γ_i for general D (one might write it $\Gamma_i^{(D)}$) is more complicated and is not discussed here, and we retain the simple notation Γ_i for the source. In Sections 2 to 4 of the text our results are given for three dimensions and their counterparts for D dimensions are given in Appendix II. In Sections 5 to 7 only the D-dimensional results are given, and they occur in the main text.

If the electron-hole pairs are collected with unit efficiency, then the current density is q multiplied by the number flux of those photons whose energies exceed E_g. It is the maximum current which can be expected and it is

$$q\Gamma\int_{y_g}^{\infty} f(y)\,g(y)\,dy\left(f(y) \equiv \frac{1}{e^y-1}, y \equiv \frac{E}{kT_p}\right)$$

Here $g(x)\,dx$ is the number flux of photons in general and is not necessarily $g_D(x)\,dx$. If the "surroundings" can be simulated by black-body radiators at temperatures $T_1(\equiv T_p), T_2, ..., T_n$, we shall use T_s to be our normalising temperature, $x \equiv E/kT_s$, so that

$$I(V) = q\int_{E_g}^{\infty}\left\{\sum_{i=1}^{n}\frac{\Gamma_i}{e^{E/kT_i}-1} - \frac{\gamma}{e^{(x-v)}-1}\right\}\tilde{g}(E)\,dE\left(v \equiv \frac{qV}{kT_s}\right) \qquad (4)$$

Re-radiation through radiative recombination from the converter at voltage V has been allowed for with γ, a geometrical correction factor, which takes into account the surface areas and solid angles that actually participate in the radiation output. The normal case when one face of the solar cell radiates into a solid angle of 2π is covered by the choice $\gamma = 1$. The case $\gamma = 2$ covers the two-sided re-radiation (the "bifacial" cell). The outgoing radiation may also be confined to a small solid angle by the use of a micromachined surface. This corresponds to $\gamma << 1$. For details, see [6, 7]. [One could regard the sum in (4) as a formal expansion of the photon flux in terms of black-body radiators, with the possibility of negative Γ_i.]

Using $t_i \equiv T_s/T_i$, $\tilde{g}_D(E)\,dE = g_D(x)\,dx = A_D(kT)^D x^{(D-1)}\,dx$, $x_g \equiv E_g/kT_s$ in (4),

$$I(V) = qA_3 \left\{ \sum_{i=1}^{n} \int_{x_g t_i}^{\infty} \frac{(kT_i)^3 \Gamma_i x^2 dx}{e^x - 1} - \gamma \int_{x_g}^{\infty} \frac{(kT_s)^3 x^2 dx}{e^{(x-v)} - 1} \right\} \tag{5}$$

The work flux output is $VI(V) \equiv W$. For an approximate maximum in the sense investigated in [2], the condition $\{\partial\,[VI(V)]/\partial\,E_g\}_V = 0$, is used, i.e.

$$\left[\frac{\partial I(V)}{\partial x_g} \right]_V = 0 \tag{6}$$

This is approximate since $(\partial W/\partial V)_{E_g} = 0$ has not been used. One obtains

$$\sum_{i=1}^{n} \frac{(kT_i)^3 \Gamma_i (x_g t_i)^2}{e^{x_g t_i} - 1} t_i = \frac{\gamma (kT_s)^3 x_g^2}{e^{(x_g - v)} - 1} \tag{7}$$

One can write this condition as giving an optimal value of v, given E_g, or as an optimal value of E_g, given v:

$$\sum_{i=1}^{n} \frac{\Gamma_i}{e^{E_g/kT_i} - 1} = \frac{\gamma}{e^{(E_g - qV_{app})/kT_s} - 1} \tag{8}$$

or

$$\sum_{i=1}^{n} \frac{\Gamma_i}{e^{E_{g,app}/kT_i} - 1} = \frac{\gamma}{e^{(E_{g,app} - qV)/kT_s} - 1} \tag{9}$$

This has the equivalent forms

$$E_{g,app} = qV + kT_s z\,(E_{g,app}) \tag{10}$$

or

$$qV_{app} = E_g - kT_s z\,(E_g) \tag{11}$$

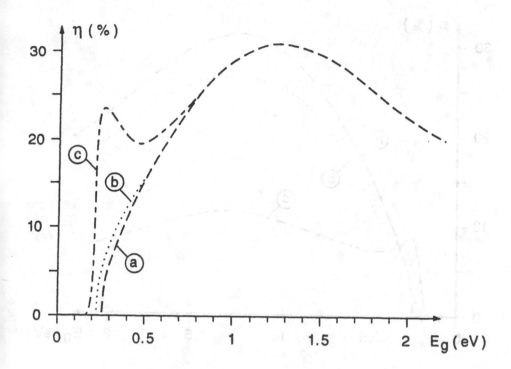

Figure 1.   Efficiency of a solar cell, according to the present work, subject to the effect of surroundings and sources 1 and 2. The efficiency is defined as the ratio of electric output power divided by the input power of source 1. $T_p \equiv T_1 = 6000K, T_2 = 800K, T_s \equiv T_3 = 300K, \Gamma_p \equiv \Gamma_1 = 2 \times 10^{-5}, \Gamma_s \equiv \Gamma_3 = 1 - \Gamma_1 - \Gamma_2.$ (a) $\Gamma_2 = 0.001, C = 50, d = 7.07$; (b) $\Gamma_2 = 0.01, C = 500, d = 22.36$; (c) $\Gamma_2 = 0.1, C = 5000, d = 70.71$.

where

$$z(E_g) \equiv \ln \left[1 + \frac{\gamma}{\sum_{i=1}^{n} \Gamma_i / (e^{E_g/kT_i} - 1)} \right] \tag{12}$$

There may of course be several efficiency maxima, as seen in figs. 1 and 2.

Observe that if the temperature of one source is increased, or a Γ_i is increased, then z is decreased: the approximately optimum energy gap is thus decreased and the approximately optimum voltage is increased. An increase in Γ_i corresponds to an increase of the apparent solid angle of source i , obtained through an increase of either the spatial extension of the source or of its optical concentration factor.

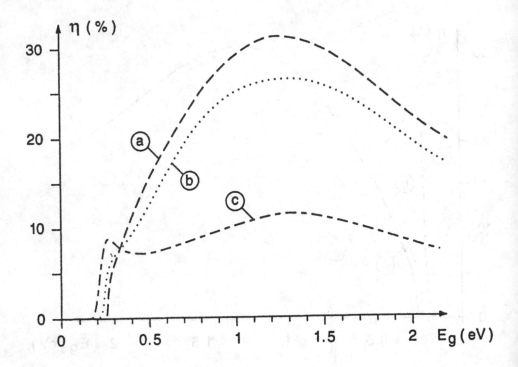

Figure 2. As Fig. 1, but with efficiency defined as the output power divided by the sum of the input powers from sources 1 and 2.

Analogous effects are produced if an additional source of temperature T_{n+1} is introduced while other things are kept the same. In that case let us put

$$A \equiv \sum_{i=1}^{n} \frac{\Gamma_i}{e^{E_i/kT_i} - 1}, C \equiv \sum_{i=1}^{n} \frac{\Gamma_{n+1}}{e^{E_{n+1}/kT_{n+1}} - 1} \qquad (13)$$

One then finds

$$\frac{E_{g,app,n} - E_{g,app,n+1}}{kT_s} = \frac{q(V_{app,n+1} - V_{app,n})}{kT_s} = \ln \frac{1 + \frac{\gamma}{A}}{1 + \frac{\gamma}{A+C}} \qquad (14)$$

If $C/A \ll 1$ the right-hand side simplifies to

$$\frac{\gamma C}{A(A + \gamma)} \qquad (15)$$

At any rate if all the Γ_i are positive, then (14) is a positive quantity so that the approximately optimal E_g is decreased by the additional pump.

3. Special Cases

One merit of the above formulation is that it yields readily several suggestive special cases:

3.1.

$\underline{n = 1}$

Here $\Gamma_1 = \Gamma_p$ corresponds to the sun as the only effective pump. The cell at temperature T_s is in these cases locked by conduction to the temperature of the surroundings, which are radiatively *uncoupled* from the cell. Otherwise the surroundings would act as a second pump.

Suppose $\Gamma_p = \gamma$. This will normally correspond to a one-sided cell ($\gamma = 1$) combined with an ideal concentrator which gives high concentration ($\Gamma_p = 1$). This is the full 2π-geometry noted in connection with (1), and it also shows that γ for the cell plays a somewhat similar role to the Γ_i for the pumps. The case $\Gamma_p = \gamma$ could also correspond to little or no light concentration ($\Gamma_p << 1$) combined with an ideal micromachining of the cell surface, as suggested in [5, 6]. For the case $n = 1$ (12) yields

$$z\left(E_g\right) = \ln\left[1 + e^{E_g/kT_p} - 1\right] = \frac{E_g}{kT_p} \qquad (16)$$

so that from (11)

$$qV_{app} = E_g\left(1 - T_s/T_p\right) \qquad (17)$$

This is the Carnot-type result already noted on p. 6419 of [3]. From (10) one finds similarly equation (1) noted above.

In other cases of $n = 1$ one finds

$$qV_{app,1} = E_g - kT_s\ln\left[1 + \frac{\gamma}{\Gamma_p}\left(e^{E_g/kT_p} - 1\right)\right] \qquad (18)$$

and $E_{g,app,1}$ is found similarly. In fact

$$E_g - E_{g,app,1} = qV_{app,1} - qV \qquad (19)$$

3.2.

$\underline{n = 2}$ ($\Gamma_1 = \Gamma_p; \Gamma_2 = \Gamma_s$ corresponds to ambient radiation.)

In this case the surroundings are coupled radiatively to the cell and

$$qV_{app2} = E_g - kT_s\ln\left[1 + \frac{\gamma}{\Gamma_p(e^{E_g/kT_p} - 1)^{-1} + \Gamma_s(e^{E_g/kT_s} - 1)^{-1}}\right] \qquad (20)$$

One may of course have $\Gamma_s = 1 - \Gamma_p$, but this is not essential if certain areas of the sky are regarded as making a negligible contribution. The approximation to the maximum in these calculations has been shown to be very good (within 1 %) [2, 8].

Similarly, if one looks for an approximately optimum E_g,

$$E_g - E_{g,app,2} = qV_{app,2} - qV \tag{21}$$

4. The Converter Efficiency

To evaluate an efficiency (η) of energy conversion we need, as the denominator, the energy flux from sources 1 to n, adjusted to vanish if all sources are at the temperature T_s of the surroundings:

$$P_3 \equiv (l/2)\sigma \sum_{i=1}^{n} \Gamma_i T_i^4 \left[1 - \left(\frac{T_s}{T_i}\right)^4\right] \equiv \frac{(l/2)\,\pi^5}{15c^2h^3} \sum_{i=1}^{n} \Gamma_i (kT_i)^4 \left[1 - \left(\frac{T_s}{T_i}\right)^4\right] \tag{22}$$

where σ is Stefan's constant.

For reasonably high pump temperatures T_i, the correction factor in square brackets at the end of equations (22) can be safely taken as unity. For low temperatures, however, for instance if $T_i < T_s$, a source can act negatively. For the efficiency we need (5) and (22) and it is convenient to define the integrals

$$F_m(u) \equiv \int_u^\infty \frac{x^m \, dx}{e^x - 1} \tag{23}$$

$$H_m(u, v) \equiv \int_u^\infty \frac{x^m \, dx}{e^{(x-v)} - 1} = \sum_{i=0}^{m} \binom{m}{i} v^i F_{m-i}(u - v)$$

$$H_2(u, v) \equiv F_2(u - v) + 2vF_1(u - v) + v^2 F_0(u - v) \tag{24}$$

One then finds for the efficiency $VI(V)/P_3$

$$\eta = \frac{qV}{kT_s} \frac{A_3(kT_s)^4}{P_3} \left\{ \sum_{i=1}^{n} \left(\frac{T_i}{T_s}\right)^3 \Gamma_i F_2\left(\frac{E_g}{kT_i}\right) - \gamma H_2\left(\frac{E_g}{kT_i}, \frac{qV}{kT_s}\right) \right\} \tag{25}$$

Suppose now that the main pumping action is due to the sources $i = 1, 2, ..., n-1$ while the nth source is due to radiation input from the ambient. If we assume that

$$\Gamma_n = 1 - \sum_{i=1}^{n-1} \Gamma_i \tag{26}$$

then

$$\eta = B \left\{ \sum_{i=1}^{n-1} \Gamma_i \left[\left(\frac{T_i}{T_s} \right)^3 F_2 \left(\frac{E_g}{kT_i} \right) - \left(\frac{T_n}{T_s} \right)^3 F_2 \left(\frac{E_g}{kT_n} \right) \right] \right.$$
$$\left. + \left(\frac{T_n}{T_s} \right)^3 F_2 \left(\frac{E_g}{kT_n} \right) - \frac{\gamma}{\rho(V)} H_2 \left(\frac{E_g}{kT_s}, \frac{qV}{kT_s} \right) \right\} \tag{27}$$

where from (25)

$$B \equiv \frac{qV}{kT_s} \frac{A_3(kT_s)^4}{P_3} \tag{28}$$

$$\equiv \frac{qV/kT_s}{\pi^4/15} \left\{ \sum_{i=1}^{n} \Gamma_i \left(\frac{T_i}{T_s} \right)^4 \left[1 - \left(\frac{T_s}{T_i} \right)^4 \right] \right\}^{-1} \tag{29}$$

Here $\rho(V)$ is a radiative recombination efficiency which converts the last term in (27) from being due to radiative recombination to covering also non-radiative recombination, as discussed in [2] and [8]. Non-radiative recombination must clearly occur, and this drags down the efficiency; but this effect will not be considered here. An example for two sources plus ambient is given in figs. 1 and 2. The symbol C denotes the light concentration factor. Bearing in mind that the effective diameter of source 2 is given by $d = \sqrt{\Gamma_2/\Gamma_p}$ times the apparent diameter of the sun, one finds the three corresponding values of d, which are also given.

In fig. 1 the power supplied by the second source is not included in the denominator of the efficiency, while it is so included in fig. 2, thus reducing all efficiencies shown in this figure. Which of these two definitions is most suitable will depend on the individual situation under consideration. The first source represents the sun (T_p) and the surroundings are radiatively coupled to the cell (T_3, Γ_3). The second source is chosen rather arbitrarily to illustrate its effect. If it subtends a sufficient angle, the efficiencies may have a second maximum as a function of the energy gap. The Shockley-Queisser ultimate efficiency, which assumes $T_c = 0$ ([3], Section 3) is maximal at $E_g = 2.17kT$, i.e. at 0.15 eV for source 2. This is pushed to higher values by the fact that $T_c > 0$ and the solid angle, and hence the concentration, is small. The latter effect is seen for example in fig. 7 of [8]. This explains the order of magnitude of the maxima in figs. 1 and 2.

By considering the case $\Gamma_1 = \Gamma_2 = ... \Gamma_{n-1} = 0$ and therefore $I = 0$ for $V = 0$, one finds an expression for the last-term-but-one in (27). Inserting it into (27) as formulated for the general case, one finds

$$\eta = B \left\{ \sum_{i=1}^{n} \Gamma_i \left[\left(\frac{T_i}{T_s} \right)^3 F_2 \left(\frac{E_g}{kT_i} \right) - \left(\frac{T_n}{T_s} \right) F_2 \left(\frac{E_g}{kT_n} \right) \right] \right.$$

$$+\frac{\gamma}{\rho(o)}\left[H_2\left(\frac{E_g}{kT_s},0\right)-\frac{\rho(o)}{\rho(V)}H_2\left(\frac{E_g}{kT_s},\frac{qV}{kT_s}\right)\right]\right\} \tag{30}$$

One may assume as in [2], that $\rho(V)$ is independent of V. Then the case $n=2$ yields a dependence of η on ρ and $\Gamma_p=\Gamma_1$ only through the product $\rho\Gamma_p$. Also $T_2=T_s$, so that

$$\eta=\frac{(qV/kT_p)\,(T_s/T_p)^3}{\pi^4/15}\left\{\left(\frac{T_p}{T_s}\right)^3 F_2\left(\frac{E_g}{kT_p}\right)-F_2\left(\frac{E_g}{kT_s}\right)\right.$$

$$+\frac{\gamma}{\Gamma_p\rho}\left[H_2\left(\frac{E_g}{kT_s},0\right)-H_2\left(\frac{E_g}{kT_s},\frac{qV}{kT_s}\right)\right]\right\}\frac{1}{1-(T_s/T_p)^4} \tag{31}$$

By comparing (30) and (31) we see that this convenient feature is a peculiarity of the case $n=2$. Indeed, replacing γ by unity, equation (31) repeats the analogous efficiency (18) of [8] and, with a slight change of notation, equation (7) of [2]. One sees therefore that efficiency equations such as (25), (29) and (30) represent generalisations of this earlier result to multisource situations.

Let us revert to the general result (25) with a view to finding its maximum value. An approximate maximum which occurs if V is regarded as given, is obtained by substituting (10) into (25):

$$\eta_{app,max}=\frac{E_{g,app}-kT_sz\,(E_{g,app})}{P_3}(kT_s)^3 A_3$$

$$\times\left\{\sum_{i=1}^n\Gamma_i\left(\frac{T_i}{T_s}\right)^3 F_2\left(\frac{E_{g,app}}{kT_i}\right)\right.$$

$$-\gamma H_2\left(\frac{E_{g,app}}{kT_s},\frac{E_{g,app}-kT_sz\,(E_{g,app})}{kT_s}\right)\right\} \tag{32}$$

The next differentiation for the real maximum of η, namely with respect to $qV=E_{g,app}-kT_sz\,(E_{g,app})$, is in effect a differentiation with respect to $E_{g,app}$. The condition for the optimum η is then

$$\frac{d}{d\,E_{g,app}}\{(32)\}=0 \tag{33}$$

In ([3], p. 6420) the limit $T_s\to 0$ of such a condition was shown to give the same maximum as was obtained by maximising the Shockley-Queisser ultimate efficiency. The analogous result will be found here. We therefore note (for consideration in Section 5) that (33) gives

$$\frac{d}{d\,E_{g,app}}\left\{\frac{E_{g,app}\sum_{j=1}^n\Gamma_j(kT_j)^3 F_2\,(E_{g,app}/kT_j)}{(\pi^4/15)\sum_{i=1}^n\Gamma_i(kT_i)^4\left[1-(T_s/T_i)^4\right]}\right\}=0 \tag{34}$$

in the limit $T_s \to 0$.

5. The Shockley-Queisser "Ultimate Efficiency"

This well-known approach depends on neglecting in (5) the re-radiation from the cell, corresponding to $\gamma = 0$, and to assume that for $E > E_g$ each photon of energy E contributes E_g to the energy. The output energy flux yields an efficiency when one divides by (23) to yield [$\zeta(x)$ is the Riemann zeta function]

$$\eta_{ult} = \frac{E_g \sum_{j=1}^{n} \Gamma_j (kT_j)^D F_{D-1}(E_g/kT_j)}{\Gamma(D+1)\zeta(D+1)\sum_{i=1}^{n} \Gamma_i (kT_i)^{(D+1)}\left[1 - (T_s/T_i)^{(D+1)}\right]} \tag{35}$$

Maximising with respect to E_g gives the condition on E_g, $E_g = E_{g0}$ say, for a maximum of η_{ult} as

$$\sum_{i=1}^{n} \Gamma_i (kT_i)^D F_{D-1}(E_{g0}/kT_i) = E_{g0}^D \sum_{i=1}^{n} \frac{\Gamma_i}{e^{(E_{g0}/kT_i)} - 1} \tag{36}$$

This generalises eq. (15) of ref. [2] to n sources and also by allowing for the Γ_i factors. For $n = 1$ and, writing T_p for T_1, the condition is after cancelling $\Gamma_1 = \Gamma_p$,

$$F_{D-1}(E_{g0}/kT_p) = \frac{(E_{g0}/kT_p)^D}{e^{(E_{g0}/kT_p)} - 1} \left(\begin{matrix} n=1 \\ \text{all } \Gamma_p \end{matrix} \right) \tag{37}$$

This is the earlier result. Using (36) in (35),

$$\eta_{ult,max} = \frac{E_{g0}^{D+1} \sum_{j=1}^{n} \Gamma_j / \left[e^{(E_{g0}/kT_j)} - 1\right]}{\Gamma(D+1)\zeta(D+1)\sum_{i=1}^{n} \Gamma_i (kT_i)^{(D+1)}\left[1 - (T_s/T_i)^{(D+1)}\right]} \tag{38}$$

with E_{g0} given by (37). For $n = 1$ (38) is

$$\eta_{ult,max} = \frac{1}{\Gamma(D+1)\zeta(D+1)} \frac{(E_{g0}/kT_p)^{(D+1)}}{e^{(E_{g0}/kT_p)} - 1} \frac{1}{\left[1 - (T_s/T_p)^{(D+1)}\right]} \tag{39}$$

Equations (38) and (39) are generalisations of equation (5) of ref. [3] for concentration factors Γ_i, multiple sources n and general dimension D.

Observe in conclusion that if E_g in (35) is replaced by $E_{g,app}$ the condition for a maximum of η_{ult} with respect to E_g is the same as (34), confirming the equivalence between the "exact" approach of Section 4 and the

Shockley-Queisser approach of this section in the limit $T_s \to 0$. The word "exact" is put in inverted commas, as the theory is exact only within our model, which assumes that all recombination is radiative - this is often called the "detailed balance" model.

6. Approximately Optimal Efficiency: A Special Case

In this section an alternative discussion of the approximate optimum efficiency is given which is suitable for certain special cases. We start with (5) to obtain the approximately maximal work flux, using (24), as

$$
W\left(V_{app}\right) = qV_{app}A_D \left\{ \sum_{i=1}^{n} (kT_i)^D \Gamma_i F_{D-1}\left(\frac{E_g}{kT_i}\right) \right.
$$
$$
\left. -\gamma(kT_s)^D \int_{z(E_g)}^{\infty} \frac{(y+v_{app})^{(D-1)}dy}{e^y - 1} \right\} \tag{40}
$$

where we have changed the variable to $y \equiv x - v_{app}$ using (11) and $v_{app} \equiv qV_{app}/kT_s$. Dividing by (23), we obtain

$$
\eta_{app,max} = qV_{app} \frac{\sum\limits_{i=1}^{n} \Gamma_i (kT_i)^D F_{D-1}\left(\frac{E_g}{kT_i}\right) - \gamma(kT_s)^D \int_{z(E_g)}^{\infty} \frac{(x+v_{app})^{(D-1)}dx}{e^x-1}}{\Gamma(D+1)\zeta(D+1)\sum\limits_{j=1}^{n} \Gamma_j (kT_j)^{(D+1)}\left[1-(T_s/T_j)^{(D+1)}\right]}
$$
$$\tag{41}$$

This approach appears most suitable for simple cases. We consider here $D = 3, n = \gamma = 1, \Gamma_1 \equiv \Gamma_p = 1$ and abbreviate

$$
y_g \equiv E_g/kT_p, \quad v_{app} \equiv (E_g/kT_s)(1-t), \quad t \equiv T_s/T_p
$$

Then (41) becomes, omitting now the factors $\left[1-(T_s/T_p)^4\right]$,

$$
\eta_{app\,max} = \frac{15}{\pi^4}\int_{y_g}^{\infty} \frac{\phi(x,t)\,dx}{e^x - 1} \tag{42}
$$

where

$$
\phi(x,t) \equiv y_g(1-t)^2\left\{x^2 + t(x-y_g)[(1+t)x + (1-t)y_g]\right\} \tag{43}
$$

The Shockley-Queisser ultimate efficiency for our special case is well-known, and is of course readily obtainable from (35) to be

$$
\eta_{ult} \equiv \frac{15}{\pi^4}y_g \int_{y_g}^{\infty} \frac{x^2\,dx}{e^x - 1} \tag{44}
$$

One can obtain an inequality for (43) which involves (44) by writing (43) in two alternative ways:

$$\frac{\phi(x,t)}{y_g} = (1-t)^2(x^2+u), u \equiv t(x-y_g)[(1+t)x+(1-t)y_g] \quad (45)$$

$$\frac{\phi(x,t)}{y_g} = (1-t)^2(1+t+t^2)x^2 - t(1-t)^2\left[2ty_g x + (1-t)y_g^2\right] \quad (46)$$

which will be used to obtain respectively a lower and upper limit of (43).

Since $x \geq y_g$ in (42), $u \geq 0$ in (45) and, on omitting it, one finds something related to (44) which is too small. In fact,

$$(1-t)^2\eta_{ult} \leq \eta_{app\,max} \quad (47)$$

Next, replacing x by y_g in the last bracket of (46), one subtracts too little, and so now obtains an upper limit

$$\eta_{app\,max} \leq (1-t)^2(1+t+t^2)\eta_{ult}$$
$$+\frac{15}{\pi^4}t(1-t)^2 y_g^3(1+t)\ln(1-e^{-y_g}) \quad (48)$$

For $T_p \sim 6000K$, $E_g = 1.12eV$, $T_s/T_p = 1/20$ one finds

$$\eta_{ult} = 0.41 \quad \text{and} \quad 0.37 \leq \eta_{app\,max} \leq 0.38 \quad (49)$$

This procedure may be of wider interest, as it in effect yields analytical expressions for the efficiency (42).

7. The Solar Cell Equation

Recall first some standard results, see also [8] and [9]. The simplest form of the solar cell equation is

$$I(T_p; T_s; V) = I_o(T_s)\left[e^{(qV/kT_s)} - 1\right] - I_L(T_p) \quad (50)$$

where I_o is the reverse saturation current density and I_L is the light-induced current density due to a pump at black-body temperature T_p. The short-circuit current density and the open-circuit voltage satisfy

$$I_{sc}(T_p) = -I_L(T_p) \quad (51)$$

$$e^{(qV_{oc}/kT_s)} = 1 - I_{sc}(T_p)/I_o(T_s) \quad (52)$$

from the condition $V = 0$ and $I = 0$ respectively. It follows that V_{oc} is a function of T_p. Hence

$$I\left(T_p; T_s; V\right) = I_o\left(T_s\right)\left[e^{\left(qV/kT_s\right)} - 1\right] + I_{sc}\left(T_p\right)$$

and also

$$I_{sc}\left(T_p\right)/I_o\left(T_s\right) = 1 - e^{\left(qV_{oc}\left(T_p\right)/kT_s\right)}$$

Hence

$$\frac{I\left(T_p; T_s; V\right)}{I_{sc}\left(T_p\right)} = 1 + \frac{e^{\left(qV/kT_s\right)} - 1}{1 - e^{\left(qV_{oc}\left(T_p\right)/kT_s\right)}} \tag{53}$$

$$= \frac{w_{oc}\left(T_p; T_s\right) - w\left(T_s; V\right)}{w_{oc}\left(T_p; T_s\right) - 1} \tag{54}$$

where we have used the notation

$$w\left(T_s; V\right) \equiv e^{\left(qV/kT_s\right)} \tag{55}$$

and

$$w\left(T_s; V_{oc}\right) \equiv w_{oc}\left(T_p; T_s\right) \tag{56}$$

The form (54) of writing the solar cell equation goes back to 1975 [10] and is convenient since it shows at once that $I = 0$ for $V = V_{oc}$ and that $I = I_{sc}$ for $V = 0$.

To derive a generalised solar cell equation in the form (54), the incident and emitted photon flux are denoted by

$$\left.\begin{array}{l} \Lambda_i\left(T_1, ..., T_n\right) \equiv A_D \sum_{j=1}^{n}\left(kT_j\right)^D \Gamma_j F_{D-1}\left(E_g/kT_j\right) \\[2mm] \Lambda_e\left(T_s, V\right) \equiv \gamma\left(kT_s\right)^D A_D H_{D-1}\left(E_g/kT_s, qV/kT_s\right) \end{array}\right\} \tag{57}$$

so that

$$I\left(T_1, T_2, ..., T_n; T_s; V\right) = -q\left[\Lambda_i\left(T_1, ..., T_n\right) - \Lambda_e\left(T_s, V\right)\right]$$

The negative sign arises because the current flow is against the direction of the incident light. Since $I\left(T_1, ..., T_n; T_s; V_{oc}\right) = 0$,

$$I\left(T_1, ..., T_n; T_s; V\right) = q\left[\Lambda_e\left(T_s, V\right) - \Lambda_e\left(T_s, V_{oc}\right)\right] \tag{58}$$

So again, as in (52), one sees that V_{oc} depends on $T_1, ..., T_n$. Also for short-circuit

$$I_{sc}\left(T_1, ..., T_n; T_s; 0\right) = q\left[\Lambda_e\left(T_s, 0\right) - \Lambda_e\left(T_s, V_{oc}\right)\right] \tag{59}$$

Division of (58) by (59) now yields

$$\frac{I\left(T_1, ..., T_n; T_s; V\right)}{I_{sc}\left(T_1, ..., T_n; T_s; 0\right)} = \frac{w_{oc}\left(T_p; T_s\right) - w\left(T_s; V\right)}{w_{oc}\left(T_p; T_s\right) - 1} \tag{60}$$

where with (56) (as before)

$$w \equiv w\left(T_s; V\right) \equiv \Lambda_e\left(T_s, V\right) / \Lambda_e\left(T_s, 0\right) \tag{61}$$

The result (60) is formally the same as (54), except that w has now a more general interpretation.

It is desirable to show that (55) is obtainable from (61) as an approximation. For this one assumes the reasonable inequality

$$0 < e^{(qV - E_g)/kT_s} < 1 \tag{62}$$

The denominator in the integral H_{D-1} can then be expanded in terms of integrals of the form

$$J_{D-1,r}\left(E_g/kT_s\right) \equiv \int_{E_g/kT_s}^{\infty} x^{(D-1)} e^{-rx} d x \tag{63}$$

One finds for (61)

$$w\left(T_s; V\right) = \frac{H_{D-1}\left(E_g/kT_s, qV/kT_s\right)}{H_{D-1}\left(E_g/kT_s, 0\right)} = \frac{\sum_{r=1}^{\infty} e^{rqV/kT_s} J_{D-1,r}\left(E_g/kT_o\right)}{\sum_{r=1}^{\infty} J_{D-1,r}\left(E_g/kT_s\right)} \tag{64}$$

As the integrals J become rapidly smaller as r increases, it suffices to take, as an approximation, only the terms with $r = 1$ in (64). Hence one obtains (55) as required. A similar argument goes through for a general density of states [11] (p.529).

8. Entropy Generation in a Solar Cell

The efficiency of a solar cell is of course limited by the Carnot formula

$$\eta \leq \eta_C \equiv 1 - T_s/T_p$$

A lower optimum efficiency results if one takes into account the fact that the cell radiates back to the sun and that entropy is generated in the cell (at a rate $\dot{S}_g$, say). Then [12]

$$\eta \leq \eta_1 \equiv 1 - \left(\frac{T_c}{T_p}\right)^4 - \frac{4}{3}\frac{T_s}{T_p}\left[1 - \left(\frac{T_c}{T_p}\right)^3\right] - \frac{T_s \dot{S}_g}{\phi_p}$$

$$\leq 1 - \left(\frac{T_c}{T_p}\right)^4 - \frac{4}{3}\frac{T_s}{T_p}\left[1 - \left(\frac{T_c}{T_p}\right)^3\right] \equiv \eta^\cdot$$

Where ϕ_p is the energy flux provided by the pump. For the definition of η^* we have neglected the entropy generation.

If one assumes that cell and surroundings are in equilibrium ($T_s = T_c$) one finds a yet higher efficiency limit

$$\eta \leq \eta^* \leq \eta^{**} \equiv 1 - \frac{4}{3}\frac{T_s}{T_p} + \frac{1}{3}\left(\frac{T_s}{T_p}\right)^4 \leq \eta_C$$

This is, however, still smaller than the Carnot limit. That η^{**} exceeds η^* can be seen from

$$\eta^{**} - \eta^* = \left(\frac{T_c - T_s}{T_p}\right)^2 \left[\left(\frac{T_c}{T_p}\right)^2 + \frac{2}{3}\frac{T_s T_c}{T_p^2} + \frac{1}{3}\left(\frac{T_s}{T_p}\right)^2\right]$$

One could simulate the system "converter or cell + pump + surroundings" by a radiator which absorbs the pump or solar radiation and radiates itself so as to produce a steady state. The re-radiated heat can then act as the hot reservoir (at T_c) of a Carnot engine, entropy generation having occurred in the radiator. The cold reservoir is here represented by the surroundings. This arrangement incorporates automatically an entropy production term in the efficiency which is a product of two factors, one for each step:

$$\eta_2 \equiv \left[1 - \left(\frac{T_c}{T_p}\right)^4\right]\left[1 - \frac{T_s}{T_c}\right]$$

The assumption that $\eta_1 = \eta_2$ thus yields an expression for the entropy generation in the solar cell:

$$\frac{\dot{S}_g}{\sigma T_p^3} = \frac{1}{3}\left(\frac{T_c}{T_p}\right)^3 + \frac{T_p}{T_c} - \frac{4}{3}$$

Here σ is again Stefan's constant. One sees that $\dot{S}_g \geq 0$ and this result gains strength from an independent derivation (which does not require $\eta_1 = \eta_2$) [13].

9. Conclusions

Previous related work, mainly contained in references [2] and [3], has been generalised to the case of n black-body pumping sources acting on a solar cell in a D-dimensional setting. The Carnot-type relation for the approximately optimal energy gap is generalised [see eq.(10)]. The close similarity between the exact optimum efficiency and that associated with Shockley-Queisser which was found before (in [2]) carries over into the generalised theory [cf. eq. (34) and Section 5 and the inequality (48)]. The well-known solar cell equation is also extended for the new generalised framework (see Section 7). Entropy generation in a solar cell has also been discussed (Section 8).

A brief account of the work in Sections 1-5 for $D = 3$ is also available [14].

Acknowledgment

We are indebted to Dr. J.K. Liakos, Southampton and Dr. Badescu, Bucharest, for discussions and to the European Commission for support through contracts JOU2-CT93-0328, ERB CH RXCT 92 0007 and CIPA-CT92-4026.

Appendix I (relevant to Section 2) The Geometrical Factor For $D = 3$

Consider a spherical black-body source whose center lies at an angle θ_o to the normal to a receiving surface element $d\Sigma$. Then for $\theta_o = 0$ integration over the solid angle subtended by the source at $d\Sigma$ gives the following geometrical factor for the usual number, or energy, or entropy, flux

$$B = \int\int \cos\theta d\omega = \int_0^{2\pi} d\phi \int_0^\delta \cos\theta \sin\theta d\theta$$
$$= \frac{\pi}{2}(1 - \cos 2\delta) = \pi\left(\delta^2 - \frac{\delta^4}{3} + \frac{2}{45}\delta^6 + \dots\right) \qquad (65)$$

Thus the energy flux ϕ is related to the luminance K by $\phi = BK$. The cone subtended by the source at the receiver is assumed to have the half-angle δ. The solid angle is similarly

$$\Omega = \int\int d\omega = \int_0^{2\pi} d\phi \int_0^\delta \sin\theta \, d\theta = 2\pi(1 - \cos\delta)$$
$$= \pi\left(\delta^2 - \frac{\delta^4}{12} + \dots\right) \qquad (66)$$

For $\delta = \pi/2$, i.e. a cell surrounded by a hemispherical source, we have

$$\Omega = 2\pi \text{ and } B = \pi \qquad (67)$$

A geometrical factor normalised for a hemisphere is therefore

$$\Gamma \equiv B/\pi \qquad (68)$$

Applied to (4), this means that all the Γ's add up to unity, and $\gamma = 1$ as well. This ensures incidentally that when all the temperatures are equal $(T_1 = \ldots = T_n)$ and $v = 0$, then $I(0)$ does indeed vanish. For a bifacial cell, or a (topologically equivalent) spherical cell, $\gamma = 2$ and all the Γ's must add up to 2.

For $\theta_o > 0$, the implied rotation of the source relative to Σ does not affect Ω. But to ensure that the whole of the sphere is visible from $d\Sigma$ one requires

$$\theta_o + \delta \le \pi/2$$

Now B goes over into

$$B(\Omega, \theta_o) = \int \underline{n} \cdot \underline{u} d\omega \qquad (69)$$

Here $\underline{n}$ is the unit vector perpendicular to the irradiated surface element and $\underline{u}$ is the unit vector from the element to a given point A on the source. If $\underline{c}$ is a unit vector from the element to the center of the source, we can set up a Cartesian coordinate system perpendicular to $\underline{c}$ as follows. The circle at right angles to $\underline{c}$ and passing through A intersects the plane defined by $\underline{n}$ and $\underline{c}$ at a point B. Then $\underline{n}_1$ is the unit vector from the center of this circle to the point B, and $\underline{n}_2$ is perpendicular to $\underline{n}_1$ and to $\underline{c}$. Note incidentally that the rays based on $\underline{n}$ and $\underline{n}_1$ intersect at an angle $\pi/2 - \theta_o$. One finds, if the angle θ is taken with respect to the vector $\underline{c}$,

$$\underline{u} = \underline{c} \cos\theta + \underline{n}_1 \sin\theta\cos\phi + \underline{n}_2 \sin\theta\sin\phi$$

We have $\underline{n} \cdot \underline{c} = \cos\theta_o$ so that $\underline{n} \cdot \underline{n}_1 = -\sin\theta_o$. Also $\underline{n} \cdot \underline{n}_2 = 0$ and

$$B = \int_0^{2\pi} d\phi \int_0^\delta (\cos\theta_o \cos\theta - \sin\theta_o \ \cos\phi \sin\theta) \sin\theta d\theta$$

$$= \pi \cos\theta_o \sin^2\delta \qquad (70)$$

(70) can be justified by the constancy of the solid angle under rotation. It can also be obtained by using the Jacobian involved in the change of coordinates. Now, using (66), one can also write

$$B = \Omega(1 - \Omega/4\pi)\cos\theta_o \qquad (71)$$

(70) and (71) are our main results.

The $\theta_o = 0$ case is important and can also be discussed as follows ([11], p.503). If r is the radius of the spherical source and d its distance from the receiver, then for $r \ll d$ the converter intercepts a fraction

$$\Gamma = (r/d)^2 = \sin^2\delta \qquad (72)$$

of the emitted number (or emitted energy) flux of photons. Using (66) and (72)

$$\Gamma = (1 - \cos\delta)(1 + \cos\delta) = \frac{\Omega}{2\pi}\left(2 - \frac{\Omega}{2\pi}\right) = \frac{\Omega}{\pi}\left(1 - \frac{\Omega}{4\pi}\right) \tag{73}$$

This means $B = \pi\Gamma$ is consistent with (71).

For direct isotropic radiation incident on the surface Σ the energy flux is

$$\Phi = B(\Omega, \theta_o)\int_0^\infty K_\nu d\nu \tag{74}$$

where K_ν is the spectral energy radiance at the surface. For black-body radiation, for example, one find the usual result ($l = 1$ for polarised, $l = 2$ for unpolarised radiation)

$$\Phi = B\int_0^\infty K_\nu d\nu = B\int_0^\infty \frac{lh\nu^3}{c^2}\frac{d\nu}{e^{(h\nu/kT)} - 1} = \frac{\pi^4 Blk^4}{15h^3c^2}T^4$$

$$= (Bl/2\pi)\sigma T^4 = (\Gamma l/2)\sigma T^4 \tag{75}$$

where $\sigma = 2\pi^5 k^4/15h^3c^2$ is Stefan's constant.

Appendix II (relevant to Sections 2-4) Some Generalisations to D Dimensions of Equations of the Main Text

Eq. (5)

$$I(V) = qA_D\{\Sigma_{i=1}^n \int_{x_g t_i}^\infty \frac{(kT_i)^D\Gamma_i x^{(D-1)}dx}{e^x - 1}$$

$$-\gamma\int_{x_g}^\infty \frac{(kT_s)^D x^{(D-1)}dx}{e^{(x-v)} - 1}\}$$

Eq. (7)

$$\Sigma_{i=1}^n \frac{(kT_i)^D\Gamma_i(x_g t_i)^{(D-1)}}{e^{x_g t_i} - 1}t_i = \frac{\gamma(kT_s)^D x_g^{(D-1)}}{e^{(x_g - v)} - 1}$$

Eq. (22)

$$P_D = \Gamma(D+1)\zeta(D+1)A_D\Sigma_{i=1}^n\Gamma_i(kT_i)^{(D+1)}\left[1 - \left(\frac{T_s}{T_i}\right)^{(D+1)}\right]$$

Eq. (25)

$$\eta = \frac{qV}{kT_s}\frac{A_D(kT_s)^{(D+1)}}{P_D}\{\sum_{i=1}^n \left(\frac{T_i}{T_s}\right)^D\Gamma_i F_{D-1}\left(\frac{E_g}{kT_i}\right)$$

$$-\gamma H_{D-1}\left(\frac{E_g}{kT_i}, \frac{qV}{kT_s}\right)\}$$

Eq. (27)

$$\eta = B\left\{\sum_{i=1}^{n-1}\Gamma_i\left[\left(\frac{T_i}{T_s}\right)^D F_{D-1}\left(\frac{E_g}{kT_i}\right) - \left(\frac{T_n}{T_s}\right)^D F_{D-1}\left(\frac{E_g}{kT_n}\right)\right]\right.$$
$$\left. + \left(\frac{T_n}{T_s}\right)^D F_{D-1}\left(\frac{E_g}{kT_n}\right) - \frac{\gamma}{\rho(V)}H_{D-1}\left(\frac{E_g}{kT_s}, \frac{qV}{kT_s}\right)\right\}$$

Eq. (28)

$$B \equiv \frac{qV}{kT_s}\frac{A_D(kT_s)^{(D+1)}}{P_D}$$

Eq. (29)

$$B \equiv \frac{qV/kT_s}{\Gamma(D+1)\zeta(D+1)}\left\{\sum_{i=1}^{n}\Gamma_i\left(\frac{T_i}{T_s}\right)^{(D+1)}\left[1 - \left(\frac{T_s}{T_i}\right)^{(D+1)}\right]\right\}^{-1}$$

Eq. (30)

$$\eta = B\left\{\sum_{i=1}^{n}\Gamma_i\left[\left(\frac{T_i}{T_s}\right)^D F_{D-1}\left(\frac{E_g}{kT_i}\right) - \left(\frac{T_n}{T_s}\right)^D F_{D-1}\left(\frac{E_g}{kT_n}\right)\right]\right.$$
$$\left. + \frac{\gamma}{\rho(o)}\left[H_{D-1}\left(\frac{E_g}{kT_s}, 0\right) - \frac{\rho(o)}{\rho(V)}H_{D-1}\left(\frac{E_g}{kT_s}, \frac{qV}{kT_s}\right)\right]\right\}$$

Eq. (31)

$$\eta = \frac{(qV/kT_p)(T_s/T_p)^D}{\Gamma(D+1)\zeta(D+1)}\left\{\left(\frac{T_p}{T_s}\right)^D F_{D-1}\left(\frac{E_g}{kT_p}\right) - F_{D-1}\left(\frac{E_g}{kT_s}\right)\right.$$
$$\left. + \frac{\gamma}{\Gamma_p\rho}\left[H_{D-1}\left(\frac{E_g}{kT_s}, 0\right) - H_{D-1}\left(\frac{E_g}{kT_s}, \frac{qV}{kT_s}\right)\right]\right\}\frac{1}{1 - (T_s/T_p)^{(D+1)}}$$

Eq. (32)

$$\eta_{app\,max} = \frac{E_{g,app} - kT_s z(E_{g,app})}{P_D}(kT_s)^D A_D\left\{\sum_{i=1}^{n}\Gamma_i\left(\frac{T_i}{T_s}\right)^D F_{D-1}\left(\frac{E_{g,app}}{kT_i}\right)\right.$$
$$\left. - \gamma H_{D-1}\left(\frac{E_{g,app}}{kT_s}, \frac{E_{g,app} - kT_s z(E_{g,app})}{kT_s}\right)\right\}$$

Eq. (34)

$$\frac{d}{d E_{g,app}}\left\{\frac{E_{g,app}\sum_{j=1}^{n}\Gamma_j(kT_j)^D F_{D-1}(E_{g,app}/kT_j)}{\Gamma(D+1)\zeta(D+1)\sum_{i=1}^{n}\Gamma_i(kT_i)^{(D+1)}\left[1 - (T_s/T_i)^{(D+1)}\right]}\right\} = 0$$

References

1. Henderson, S.T. (1977)*Daylight and its Spectrum*, Adam Hilger, Bristol, pg. 37.
2. Baruch, P., Landsberg, P.T., Vos, A. de (1992) *11th European Photovoltaic Solar Energy Conference, Montreux, October 1992*, pp. 283-286.
3. Landsberg, P.T., Vos, A. de, Baruch, P. (1991) *J. Phys. Condensed Matter* **3**, 6415-6424.
4. Vos, A. de (1987) *Solid-State Electronics* **30**, 853- 857; Vos, A. de (1988) *J. Phys. Chem. Solids* **49**, 725-730.
5. Landsberg, P.T. and Vos, A. de (1989) *J. Phys. A* **22**, 1073-1084.
6. Araújo, G.L. and Martí , A. (1992) *11th European Photovoltaic Solar Energy Conference, Montreux, October 1992*, pp. 142-145.
7. Miñano, J.C. (1990) in A. Luque and G.L. Araújo (eds.) *Physical Limitations to Photovoltaic Energy Conversion*, Adam Hilger, Bristol, pp. 50-83.
8. Baruch, P., Vos, A. de, Landsberg, P.T. and Parrott, J.E. (1995) *Solar Energy Materials and Solar Cells* **36**, 201-222.
9. Parrott, J.E. (1986) *IEE Proceedings J* **133**, 314-318.
10. Landsberg, P.T. (1975) *Solid-State Electronics* **18**, 1043-1052.
11. Landsberg, P.T. (1991) *Advances in Thermodynamics* **3**, 482-537.
12. Landsberg, P.T. (1983) *J. App. Phys.* **54**, 2841-2843.
13. Vos, A. de and Pauwels, H. (1981) *Applied Physics* **25**, 119-125.
14. Landsberg, P.T., Vos, A. de, Baruch, P. and Parrott, J.E. (1995) *Proceedings of the 13th European Photovoltaic Solar Energy Conference, Nice, October 1995*, to be published.

COUPLED TRANSPORT OF HEAT AND MASS.
THEORY AND APPLICATIONS

S. KJELSTRUP RATKJE AND B. HAFSKJOLD

Department of Physical Chemistry,
Norwegian Institute of Technology,
University of Trondheim
N-7034 Trondheim, Norway.

Abstract. One natural and two technical processes with coupled transport of heat and mass are reviewed. The processes are frost heave, freeze concentration of juice, and salt transport in carbon cathode bottoms in the aluminium electrolysis cell. Recent non-equilibrium molecular dynamics simulation results have been used to establish molecular mechanisms of coupled heat and mass transport in liquids. Linear flux-force relationships have been found for extremely large temperature gradients. Use of linear irreversible thermodynamics is therefore justified for all three practical cases considered here. Two criteria for local equilibrium are reviewed. In particular, local equilibrium was obtained in a liquid or dense gas with a temperature gradient if $l/|\vec{\nabla}T| \leq \delta T < 0.05T$, where l is the dimension of the local control volume, $\vec{\nabla}T$ is the temperature gradient, and δT is the temperature fluctuation in the control volume. This criterion is fulfilled for the subcooled solution in the process of freeze concentration. The energy transported by water moving from a subsurface water table to an ice lens in clay capillaries during frost heave is mainly the enthalpy of freezing of water, lending support to the description of frost heave as a transport process. Similarly, the separation of salts in the cathode bottom of the aluminium electrolysis cell and the formation of salt lenses (bottom heave) can be understood as a way the system reacts to a temperature gradient in order to transport energy (heat) as effectively as possible. Computer simulations have confirmed the validity of the Onsager reciprocal relations (ORR) in liquids. The application of the ORR for average phenomenological coefficients across interfaces in the systems is discussed.

1. Introduction

In this paper we review examples of coupled transport of heat and mass that we have studied by computer simulations and in relation to experiments. These coupled transport phenomena are important in many natural and technical contexts. We describe briefly the main features of the natural frost heave phenomenon [1], a practical method of freeze concentration for concentrating juice solutions [2], and one possible mechanism for bottom

197

J. S. Shiner (ed.), Entropy and Entropy Generation, 197–219.
© 1996 *Kluwer Academic Publishers.*

heave in aluminium electrolysis cells [3]. Common to these transport problems is that they include a transition from one aggregate state to another at an interface, under the influence of a temperature gradient. Through this review we therefore address the theoretical problem raised by Bornhorst and Hatsopolous [4]; whether the "concept of heat of transfer fails at the interface". We shall see that the heat of transfer is well defined in a two component system. In a one component system, the heat of transfer across an interface can be compared to the enthalpy change of the phase transition.

The theory of irreversible thermodynamics (see e.g. ref. [1]) has been used to describe the phenomena mentioned above at a macroscopic level. This theory defines the coupling of heat and mass transport in different ways; by the heat of transfer, q^*, the thermal diffusion factor, α_{ij}, or by some of the off-diagonal elements of the matrix of phenomenological coefficients. The fundamental framework equations are given in Section 2 to give a common basis for the review. We shall then see, in Section 3, that the heat of transfer plays a crucial role in the descriptions of the processes mentioned.

In order to understand coupled heat and mass transport in terms of molecular properties, obtain estimates for q^* and α_{ij}, and check some of the approximations made in the macroscopic theory, we have performed molecular dynamics calculations [5-9]. Our molecular dynamics work is reviewed in Section 4, to give the status of this knowledge and to relate the work to the practical examples described in Section 3. The validity of the basic assumptions used in Section 2 is addressed in particular.

The aim of the review is thus to make a bridge between theoretical developments and practical applications.

2. Irreversible thermodynamic theory

The entropy production rate per unit volume of a binary mixture with a temperature gradient and one independent chemical force is (see e.g. ref. [1]:

$$\theta = -\vec{J}_q \frac{\vec{\nabla} T}{T^2} - \vec{J}_1 \frac{\vec{X}}{T} \tag{1}$$

where $\vec{J}_q$ is the measurable heat flux, $\vec{\nabla} T$ is the temperature gradient, and $\vec{J}_1$ is the mass flux of component 1. The chemical potential gradients of two components in a mixture depend on each other through Gibbs-Duhem's equation. The fluxes are interdependent through the frame of reference. By elimination of one of the components, we have the following possibilities for $\vec{X}$:

$$\vec{X} = \vec{\nabla}_T(\mu_1 - \mu_2) \tag{2}$$

with barycentric frame of reference for the transport, and

$$\vec{X} = c_1 \vec{\nabla}_T \mu_1 \tag{3}$$

with the velocity of component 2 (in m s^{-1}) being the frame of reference. The chemical potential of i is μ_i, c_i is the concentration (in mol l^{-1}), and subscript T denotes constant temperature. The entropy production does not depend on the frame of reference for the fluxes. Assuming linear flux-force relationships, the heat and mass fluxes are:

$$\vec{J}_q = - l_{qq}\frac{\vec{\nabla}T}{T^2} - l_{q1}\frac{\vec{X}}{T} \tag{4}$$

$$\vec{J}_1 = - l_{1q}\frac{\vec{\nabla}T}{T^2} - l_{11}\frac{\vec{X}}{T} \tag{5}$$

where l_{ij} are phenomenological coefficients. The heat flux is not purely conductive, but there is also a contribution due to the chemical force - the Dufour effect. Similarly, the mass flux has two contributions, the Fickian diffusion and the Soret effect.

The independent phenomenological coefficients of Eqs. (4,5) follow from the flux equations and the Onsager reciprocal relation (ORR) which gives $l_{q1} = l_{1q}$. In a two-component system there is only one coefficient related to diffusion, i.e. l_{11}. This coefficient is an interdiffusion coefficient.

The diagonal coefficients l_{qq} and l_{11} can be related to the thermal conductivity and interdiffusion coefficient, respectively. By elimination of the chemical force in Eq. (5) and introduction into Eq. (4) we obtain for the thermal conductivity:

$$\lambda_{J_1=0} = \left(l_{qq} - \frac{l_{q1}l_{1q}}{l_{11}} \right)\frac{1}{T^2} \tag{6}$$

and for the heat flux:

$$\vec{J}_q = -\lambda \vec{\nabla} T + q^* \vec{J}_1 \tag{7}$$

Eq. (7) defines the heat of transfer, q^*. For a homogeneous fluid mixture, q^* may be expressed as:

$$q^* = \frac{l_{q1}}{l_{11}} \tag{8}$$

The contribution to the entropy production from the term $q^* \vec{J}_1$ will be compensated by another term because of the symmetry of the l_{ij} matrix. The contribution represents a reversible part of the process (in the sense that it is reversed if $\vec{J}_1$ is reversed). The reversible nature of $q^* \vec{J}_1$ implies that if the conductive heat flux is neglected, it must be possible to describe the effects of the heat of transfer in terms of equilibrium properties at conditions corresponding to the boundaries of the system or the control volume in consideration. This will be used in Section 3 in the analysis of frost heave, freeze concentration, and salt transport in cathode materials.

The thermal diffusion factor is defined by

$$\alpha_{12} = -\left(\frac{\vec{\nabla} \ln(w_1/w_2)}{\vec{\nabla} \ln T} \right)_{\vec{J}_1 = 0} = -\frac{T}{w_1 w_2} \left(\frac{\vec{\nabla} w_1}{\vec{\nabla} T} \right)_{\vec{J}_1 = 0} \tag{9}$$

where w_i is the weight fraction (or the mole fraction) of component i. The heat of transfer for an ideal mixture (which we shall consider here) is related to the thermal diffusion factor by

$$q^* = \frac{k_B T}{(m_1 w_2 + m_2 w_1)} \alpha_{12} \tag{10}$$

where k_B is Boltzmann's constant. These equations constitute a framework for experimental studies as well as simulations.

The heat flux for a one-component system is fundamentally different from the heat flux in a mixture, in that there is no heat of transfer or thermal diffusion defined as in Eqs. (7) and (9). We consider instead the enthalpy flux. Enthalpy is transported in a flowing or diffusing one-component fluid in two ways (neglecting radiation), by conduction and convection:

$$\vec{J}_H = \vec{J}_q + H\vec{J}_1 \tag{11}$$

where the heat conduction is expressed by:

$$\vec{J}_q = -\lambda \vec{\nabla} T \tag{12}$$

and the molar flux is

$$\vec{J}_1 = n\vec{v} \tag{13}$$

Here, H is the molar enthalpy, n is the molar density, and $\vec{v}$ is the center-of-mass velocity. A practical frame of reference for $\vec{v}$ is the container of the system. The reference state of H must be chosen, and the enthalpy flux depends on the reference state for the energy as well as on the frame of reference. The heat flux does not depend on any of these references.

The amount of energy transported with e.g. a flowing vapour may be substantial. Consider therefore the enthalpy flux at isothermal conditions:

$$\left(\vec{J}_H\right)_{\vec{\nabla} T=0} = H\vec{J}_1 \tag{14}$$

In systems where the pressure gradient is small, we may neglect the macroscopic kinetic energy (center-of-mass kinetic energy), and the enthalpy is to a good approximation conserved. This implies that the enthalpy flux is constant in space. The mass flux is also constant. Comparing the heat flux in two points in space, (a) and (b), the molar enthalpy appears as a source of heat:

$$\vec{J}_q^{(a)} = \vec{J}_q^{(b)} + (H^{(b)} - H^{(a)})\vec{J}_1 \tag{15}$$

We see that the form of Eq. (15) is similar to that of Eq. (7), with the interpretation that $q^* = \Delta H = H^{(b)} - H^{(a)}$. However, Eq. (7) refers to a single point in space, whereas Eq. (15) refers to the difference between two points. Moreover, the heat flux in Eq. (7) is partly transported heat, but the heat fluxes in Eq. (15) are purely conductive. In Eq. (15) the heat of transfer therefore represents a source term in the heat balance. In particular, the measurable heat flux through an interface will experience a source term, $\Delta H \vec{J}_1$, where ΔH is the change in molar enthalpy on the phase transition, when energy is conserved in the process.

3. Examples of Coupled Transport of Heat and Mass

3.1. FROST HEAVE

Frost heave occurs when there are temperature gradients across a soil which contains liquid water below 0 °C. Frost heave is characterized by the formation and growth of bulk ice lenses in the soil, and the development of a hydrostatic pressure on the ice. The transport of liquid water through soil capillaries from a high temperature to a low temperature, where it is allowed to freeze, can be seen as an efficient way of energy transport, not only by ordinary Fourier type conduction, but also by moving a latent heat of freezing, Eq. (15). The process is illustrated in Fig. 1.

Frost heave was compared to thermal osmosis by Førland et al. [1]. When the chemical composition of the soil does not vary, the system can be regarded as a limiting case of a two- component system. The presence of an interface to which water can flow and freeze, allows the development of a pressure difference over a certain distance Δx, thus the name thermal osmosis. The two driving forces in the system are the geothermal gradient and the pressure gradient, the latter given by

$$\vec{X} = V_1 \vec{\nabla} p \tag{16}$$

where V_1 is the partial molar volume of component 1. We consider a one-dimensional transport of water over Δx, encompassing the interface. Following the original literature, we omit vector signs, replace gradients by differences, use average (over Δx) transport coefficients, $\bar{l}_{ij} = (1/\Delta x)\int l_{ij}(x)dx$ and assume that $\bar{l}_{q1} = \bar{l}_{1q}$ is valid. This implies that we consider the transport as taking place between different equilibrium states. The water transport will contribute to the enthalpy transport by approximately

$$\left(J_q\right)_{\Delta T=0} = \frac{\bar{l}_{q1}}{\bar{l}_{11}} J_w = \Delta_f H_w J_w \tag{17}$$

where J_w and $\Delta_f H_w$ are the water flux and heat of fusion of water, respectively. Conservation of enthalpy was then assumed for the flow. By introducing $X = V_{ice} \Delta p$, $\bar{l}_{q1} = \bar{l}_{1q}$, and $\bar{l}_{q1} = \bar{l}_{11}\Delta_f H$ into Eq. (5) we obtain:

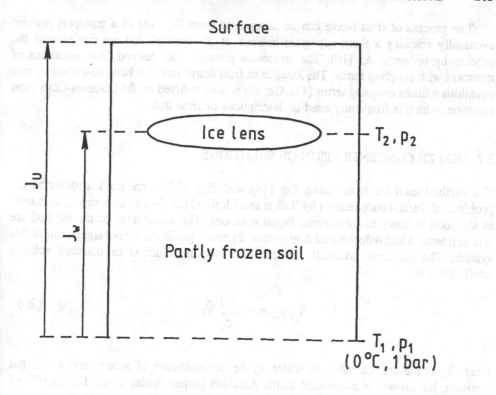

Figure 1. Schematic illustration of the heat and mass transport during frost heave. The ice lens is formed at a temperature below 0 °C, and by growing, it builds up a pressure on the soil matrix.

$$J_w = -\frac{\bar{l}_{11}}{T_f}\left(\frac{\Delta_f H_w}{T_f}\Delta T + V_{ice}\Delta p\right) \tag{18}$$

The temperature of freezing is T_f. The mass transfer coefficient $\bar{l}_{11}$ may depend largely on the low temperature layer close to the ice lense. The maximum pressure difference is obtained when J_w is zero, which is a state of balance between the thermal force and the chemical force. The experimental result for a clay material having $\Delta T < 20$ K is [10]:

$$\left(\frac{\Delta p}{\Delta T}\right)_{J_w=0} = -11.2\times10^5 \text{ Pa} \tag{19}$$

which agrees with our description by introducing known data for $\Delta_f H$, T_f, and V_{ice} into Eq. (18). This suggests that the assumptions behind Eq. (18) may be correct.

The process of frost heave can be understood from Eq. (18) as a transport process eventually reaching a thermodynamic balance. It is observed that we can reverse the process by reversing Δp [10]. The maximum pressure was derived from equations of transport with coupling terms. The kinetics of frost heave has also been described by rate equations without coupling terms [11]. Growth is not predicted by the Clausius-Clapeyron equation, which is frequently used in descriptions of frost heave.

3.2. FREEZE CONCENTRATION OF SOLUTIONS

The method used for frost heave, Eq. (18), and Eqs. (4,5) were used to describe the problem of freeze concentration by Ratkje and Flesland [2]. Freeze concentration is used in the food industry to concentrate liquid solutions. The advantages of the method are aroma retention and reduced heat degradation. Pressure gradients are not significant in this system. The chemical potential gradient of water in solutions of constant activity coefficients is:

$$\vec{\nabla}_T \mu_w = - \frac{RT}{c_w} \vec{\nabla} c_s \qquad (20)$$

where c_w is the concentration of water, c_s the concentration of solute, and R the gas constant. Ice growth was assumed in the direction perpendicular to the falling film of solution, see Fig. 2.

The flux of ice (in the x-direction) from the subcooled solution was, with the same assumptions as in Section 3.1:

$$J_{ice} = - \frac{\bar{l}_{11}}{T_f} \left(\frac{\Delta_f H_w}{T_f} \Delta T - \frac{RT}{c_w} \Delta c_s \right) \qquad (21)$$

where T_f is the freezing temperature of the solution, as given by the phase diagram. Eq. (21) was assumed valid for $T < T_f$. The average coefficient $\bar{l}_{11} = (1/\Delta x)\int l_{11}(x)dx$ was estimated from experimental data for the self diffusion coefficient of water, D_w:

$$l_{11} = \frac{D_w c_w}{RT} \qquad (22)$$

and Δx was estimated from laminar film theory. Eq. (21) predicts that the ice growth from a solution of uniform concentration is directly proportional to ΔT, the temperature difference between the bulk solution and the ice. This is in agreement with experimental results for large crystals, which can be compared to a flat surface [12].

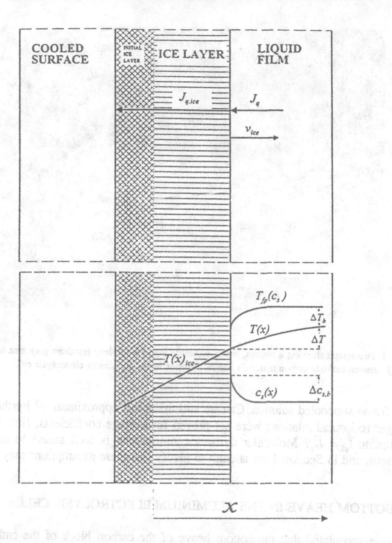

Figure 2. The transport of heat and mass from a falling film of subcooled solution. The upper part of the figure shows the heat fluxes in the two phases and the growth of the ice layer. The lower part shows the actual temperature profile, $T(x)$, the actual concentration profile of the solute, c_s, and the equilibrium freezing temperature corresponding to the local composition.

Current models [13] for freeze concentration do not predict this direct proportionality. The predicted dependence on the concentration is also at variance with other models [13], but the order of magnitude of estimated growth rates (v_{ice}) can be compared to experimental observations, 10^{-7} m s^{-1} [12].

The application of Eq. (21) is not straightforward. The basic assumption in irreversible thermodynamics is the assumption of local equilibrium, and this assumption is a priori not

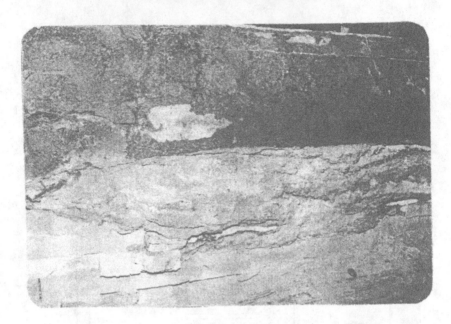

Figure 3. Photograph showing a convex, broad accumulation of a salt lens (medium grey area below the dark carbon) between cathode carbon materials and refractories in the aluminium electrolysis cell.

valid for an subcooled solution. Can we still make this approximation? Furthermore; the Onsager reciprocal relations were not proven for average coefficients. How good is the assumption $\bar{l}_{q1} = \bar{l}_{1q}$? Molecular dynamics simulations is well suited to answer such questions, and in Section 4 we attempt to justify that these assumptions may be applied.

3.3. BOTTOM HEAVE IN THE ALUMINIUM ELECTROLYSIS CELL

We have speculated that the bottom heave of the carbon block of the cathode in the aluminium electrolysis in some ways is similar to frost heave [3]. The temperature difference across the 1 m high carbon block is about 100 °C. The anthracitic carbon is porous and wetted by the electrolyte components after initial penetration of Na (metal) into the carbon [14]. The electrolyte for aluminium electrolysis contains several components, the major ones being the fluorides NaF and AlF$_3$, and both of these can be found inside the pores. Convex salt accumulations, which can be compared to a lens, are commonly observed between the carbon and the refractories, see Fig. 3.

3.3.1. *Thermal diffusion of salts*
The components of a homogeneous salt mixture will separate in a temperature gradient due to the Soret effect [15, 16]. The magnitude of heat of transfer for interdiffusion of Ag$^+$ and M$^+$ in nitrate melts, where M$^+$ is an alkali metal, varies with composition and

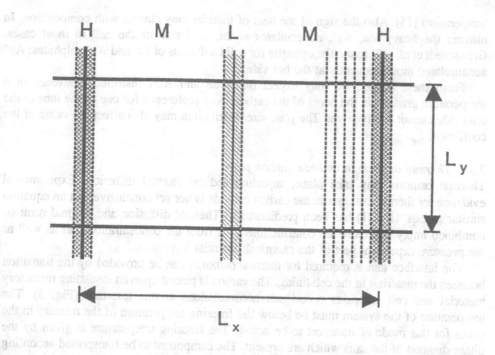

Figure 4. The NEMD cell in x-y projection. The cell has periodic boundary conditions and it is divided into 32 layers perpendicular to the x-axis (not all layers in region M are shown). Regions H and L are thermostatted to high and low temperatures, respectively.

of thermal diffusion? These questions were central in the problems described in Section 3.

Specific algorithms for NEMD, developed for coupled heat and mass transport, are given by MacGowan and Evans [21], Gillan et al. [22], Paolini and Ciccotti [23], and Vogelsang et al. [24]. We have employed a newly developed NEMD algorithm [7, 6] to coupled heat- and mass transport in a binary Lennard-Jones spline (L-J/s) isotope mixture [25]. The algorithm was chosen to represent an experimental situation with a heat source and -sink to generate a flux of enthalpy and heat through the system. It is similar to the fuzzy-wall algorithm used by Kincaid et al. [5] (see also Ashurst and Hoover, [26]).

The MD cell is not necessarily cubic and it is divided into a number of layers perpendicular to the x-axis as shown in Fig. 4. The layers form three regions, marked H, M, and L. Regions H and L have high and low temperature, respectively. The middle regions are used for computations of the local density, temperature, composition, mass fluxes, enthalpy flux, energy, and other quantities of interest. We review results for one-component flow and for an isotope mixture with mass ratio $m_1/m_2 = 10$.

The enthalpy flux $\vec{J}_H$ may be computed according to Evans and Morriss [19]:

temperature [15]. Also the sign of the heat of transfer may change with composition. In nitrates the heavy ion, Ag^+, accumulates at the cold side of the cell in most cases. Grimstvedt et al. [17] found the opposite for solid solutions of Li^+ and Ag^+ sulphates: Ag^+ accumulated most frequently at the hot side.

From these results, we may expect that NaF and AlF_3 distribute unevenly in a temperature gradient in the pores of the carbon. Any preference for one of the ions to the cold side cannot be given yet. The pore size distribution may also affect the value of the coefficient $\bar{l}_{1q}$.

3.3.2. Thermal osmosis in cathode carbon pores

Thermal osmosis may take place, superimposed on thermal diffusion. Experimental evidence for thermal osmosis in the carbon cathode is not yet conclusive, but an equation similar to Eqs. (18, 21) has been predicted [3]. Thermal diffusion and thermal osmosis combined imply that there are contributions to X from the concentration part as well as the pressure dependent part of the chemical potential.

The interface that is required for thermal osmosis, can be provided by the transition between the materials in the cell lining. The carbon is placed upon an insulating refractory material and cell autopsies reveal salt accumulations at this interface (Fig. 3). The temperature of the system must be below the freezing temperature of the mixture in the pores for this mode of transport to be active. The freezing temperature is given by the phase diagram of the salts which are present. The component to be transported according to thermal osmosis must be the component that, according to the phase diagram, precipitates as the solid for the given compositions. For some compositions investigated so far this has turned out to be NaF. For these conditions we may write:

$$J_{NaF} = - \frac{\bar{l}_{11}}{T_{liq}} \left(\frac{\Delta_f H_{NaF}}{T_{liq}} \Delta T + V_{NaF} \Delta p + \Delta \mu_{NaF}^c \right) \tag{23}$$

where $\bar{l}_{11}$ is the average transport coefficient for the distance Δx. The value of $\bar{l}_{11}$ in porous materials is probably small (diffusion in pores). A slow accumulation over the years of operation may then be expected, in accordance with observations. Chemical reactions between the penetrating salt and the refractories will complicate Eq. (23) considerably.

Additional contributions to bottom heave may come from penetration of Al (metal) and subsequent reactions of metal, bath and refractory materials [18]. Some of these reactions have a positive reaction volume.

4. Molecular Dynamics Simulations

One of the strong features of nonequilibrium molecular dynamics (NEMD) [19, 20] is the method's ability to determine a system's transport properties at the molecular scale. We shall use this feature here to analyse the following questions: What are the limits of validity of the coupled flux equations? Is $\bar{l}_{q1} = \bar{l}_{1q}$ fulfilled if we include a phase transition? How can we interpret the heat of transport across a phase boundary? What is the cause

$$\vec{J}_H = \frac{1}{V} \sum_{i \in CV} \left\{ \left[\frac{1}{2} m_i (\vec{v}_i - \vec{v})^2 + \phi_i \right] (\vec{v}_i - \vec{v}) - \frac{1}{2} \sum_{\substack{j=1 \\ j \neq i}}^{N} \left[(\vec{v}_i - \vec{v}) \cdot \vec{F}_{ij} \right] \vec{r}_{ij} \right\} \qquad (24)$$

where V is the size of the control volume (CV), m_i and $\vec{v}_i$ are the mass and velocity, respectively, of particle i, $\vec{v}$ is the barycentric velocity of the system, ϕ_i is the potential energy of particle i in the field of all the other particles, $\vec{F}_{ij}$ is the force acting on i due to j, and $\vec{r}_{ij}$ is the vector from the position of i to the position of j. In each control volume, which is an open system, N is the number of particles. This is a local, instantaneous version of the macroscopic enthalpy flux. Using Eq. (24), we interpret the enthalpy flux in terms of kinetic and potential energy contributions and by intermolecular energy transfer due to motion of a particle in the field of the other particles (the propagation of the pressure tensor).

The mass flux of component k is

$$\vec{J}_k = \frac{1}{V} \sum_{i \in CV \cup k} m_i (\vec{v}_i - \vec{v}) \qquad (25)$$

where the summation is performed over particles i of type k in the CV. The chemical force conjugate to this flux is given by Eq. (2). Similarly, we can compute the contributions to the enthalpy flux from component k by summing only over $i \in k$. The temperature gradient is in NEMD given by computations of the local temperature according to (assuming local equilibrium)

$$\frac{3}{2} NkT = \frac{1}{2} \sum_{i \in CV}^{N} m_i (\vec{v}_i - \vec{v})^2 \qquad (26)$$

where $3N$ is the number of degrees of freedom. With the fluxes, the local temperatures and the local compositions, we can examine the flux-force relationships.

4.1. CRITERIA FOR LOCAL EQUILIBRIUM IN A TEMPERATURE GRADIENT

The assumption of local equilibrium is a basic and necessary assumption in linear irreversible thermodynamics. It enables us to apply the equations of equilibrium thermodynamics, such as the Gibbs equation, to *local* volume elements in a system. This was particularly important for the modelling of freeze concentration in Section 3.2.

Local equilibrium is, however, conceptually difficult. How local is "local", and what exactly is meant by "equilibrium"? In his discussion of the local equilibrium assumption,

Kreuzer [27], considers volume elements that are small enough, so that the thermodynamic properties vary little over each element, but large enough so that each element can be treated as a macroscopic thermodynamic subsystem. In an analysis based on NEMD [9] we quantified "large enough" by a lower limit on the fluctuations in the particle number, a lower limit on a length scale comparable to the molecules' mean free path or an upper limit of the temperature fluctuations. "Small enough" was quantified by comparing the temperature difference over a volume element with its local temperature fluctuations.

Following Kreuzer, we considered a control volume of length l. The size of the control volume was determined from equilibrium fluctuation theory and the results of Tenenbaum et al. [28]; it was large enough so that the properties of the system could be precisely computed. The fluctuation in particle number was $\delta N/N \leq 0.2$ and the temperature fluctuation was $\delta T/T \leq 0.05$. Local equilibrium is maintained if $l|\vec{\nabla}T|$, where $\vec{\nabla}T$ is the temperature gradient, is smaller than the fluctuations in T [27]. A combined criterion which was derived in our NEMD calculations [9] is:

$$l\,|\vec{\nabla}T\,| \leq \delta T \tag{27a}$$

and

$$\delta T/T < 0.05 \tag{27b}$$

We derived this for an isotope mixture at reduced number density $n^* = N\sigma^3/V = 0.1$ and reduced temperature $T^* = k_B T/\varepsilon = 2.0$. Here, σ and ε are the Lennard-Jones/spline potential parameters. This criterion is in agreement with, but more specific than that proposed by other authors ($\delta T/T \ll 1$). A typical subcooling in freeze concentration is $T_f - T \approx 1$ K, with $T \approx 300$ K. If we consider the subcooling as a perturbation to the equilibrium conditions, we find that it is well within the fluctuations that the system tolerates in a state of local equilibrium. This may be a practical criterion for similar cases.

Another criterion according to Tenenbaum et al. [28] states essentially that the local density in a non-equilibrium system must be equal (within statistical uncertainties) to the equilibrium value at the same temperature and pressure. This relates "local equilibrium" to the equation of state and use of thermodynamic properties as indicators of local equilibrium. The fulfilment of this criterion is illustrated by Fig. 5 for a binary L-J/s isotope mixture at $n^* = 0.1$ and $T^* = 2.0$. Local equilibrium must also be expected at higher densities according to this criterion. This also indicates that the assumption of local equilibrium is valid in freeze concentration, where the density is higher and the temperature gradient is much smaller than in the NEMD simulation.

Additional examinations of the local equilibrium assumption were also made by Hafskjold and Ratkje [9] based on the velocity part of the Boltzmann H-function [29] and a criterion based on the Butler-Volmer equation in the overpotential theory for electric current across an interface (see e.g. ref. [30]). These criteria confirmed that local equilibrium can be assumed in the situations considered here.

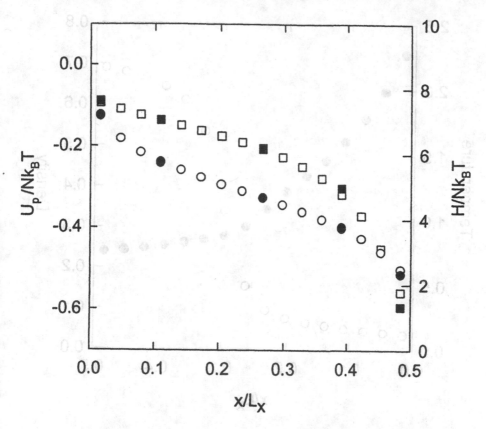

Figure 5. Equilibrium values (black symbols) and nonequilibrium results (white symbols) in a binary isotope mixture with heat flux from left to right and zero mass flux. Squares represent the potential energy, U_p/Nk_BT, and circles represent the enthalpy, H/Nk_BT.

4.2. THE VALIDITY OF ONSAGER RECIPROCAL RELATION

The ORR, applied to the systems in consideration here, states that $l_{q1} = l_{1q}$. It is central in linear irreversible thermodynamics, and the range of its validity will guide us in approximating the off-diagonal l-coefficients. We found that it is valid within uncertainty limits for the systems we have studied by NEMD simulations.

From this result and the constant gradients in the cell, we infer for the homogeneous mixture that also for a finite distance Δx, we have

$$\overline{l}_{q1} = \overline{l}_{1q} \qquad (28)$$

The question that remains to be solved is whether this is true if Δx includes an interface. This relation was applied in the derivation of Eqs. (18, 21, and 23). The reversible nature

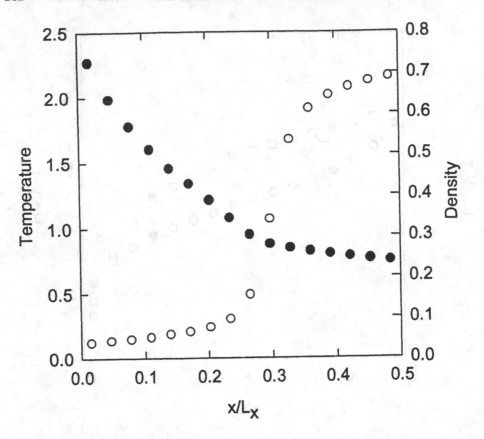

Figure 6. Density (O) and temperature (●) profiles in a one-component system. The system has separated into a vapour phase (left) and a liquid phase (right) with an interface in between.

of the phase transition, and the interpretation given by Eqs. (7,15), lead us to expect that Eq. (28) holds for the interface itself. If this is so, we may deduce that Eq. (28) also holds by extending Δx beyond the interface. In any case, considering the relative magnitude of q^* for the bulk and for the interface, we may neglect the contribution from the bulk, and assume approximate validity.

4.3. THE HEAT OF TRANSFER ACROSS INTERFACES

By appropriate choices of the overall density and the temperatures in regions H and L of the NEMD system, an interface will be established between two bulk phases [7]. For instance, Fig. 6 shows the density- and temperature profiles in a one-component system of L-J/s particles. There is no mass flux in this case. The results show approximately constant temperature gradients in each phase. The two bulk phases correspond to gas (left) and liquid (right). The contributions to the enthalpy flux, which equals the heat flux in this

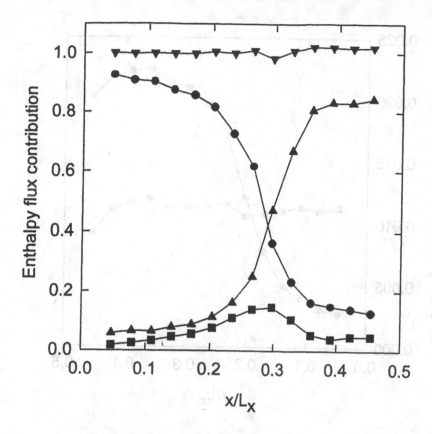

Figure 7. Contributions to $\vec{J}_H$ across a vapour/liquid interface in a one-component L-J/s system (no mass flux). The contributions are the kinetic energy flux (●), the potential energy flux (■), and the intermolecular nergy transfer (▲). The total energy flux (▼) is normalized to the stationary-state value computed from the rate of energy added and withdrawn in the regions H and L, respectively, of the NEMD cell.

case, are shown in Fig. 7.

The kinetic energy flux dominates in the vapour phase whereas the intermolecular energy flux dominates in the liquid phase. The potential energy flux is small everywhere, but has a maximum at the interface. Total enthalpy is conserved (the upper curve).

If there is an additional convective mass flux from the vapour to the liquid in the one-component system, the temperature- and density profiles remain essentially the same as in the stagnant case (Fig. 6). This justifies the introduction of $dT/dx = \Delta T/\Delta x$ in Eqs. (18, 21 and 23). The heat flux shown in Fig. 8 has a sharp increase at the interface. Analysis of the enthalpy profile confirms that the increase corresponds to $\Delta_{vap} H$ (cf. Eq. (15)). The constant energy flux is also shown.

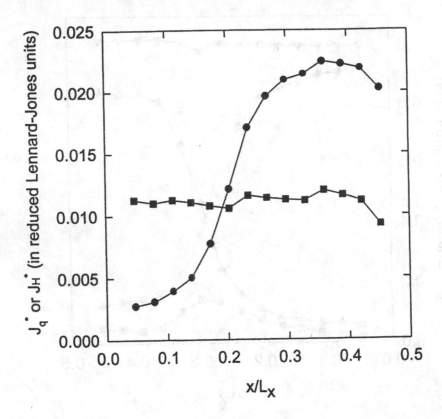

Figure 8. The heat flux (●) and enthalpy flux (■) across an interface with $\vec{J}_1 \neq 0$ (convective mass flux) in a one component L-J/s system. Note the increase in the heat flux at the interface, which is due to the liberated heat of condensation. The enthalpy flux is constant across the interface. The fluxes are here in reduced L-J/S units, but the scale for the enthalpy flux depends on the reference state of the internal energy. The fluxes are therefore not directly comparable.

4.4. THERMAL DIFFUSION IN A TWO COMPONENT MIXTURE

The sign of the thermal diffusion factor has been found experimentally to depend on the composition for some binary mixtures [31]. Kinetic theory predicts that the sign depends on the fluid density, temperature, and component mass ratio in addition to the pair potential parameters of the system [32]. Despite the experimental data and theoretical progress, there is at present no available theory that can reliably predict the sign or the values of the thermal diffusion factor or heat of transport [8].

It seems clear that isotope mixtures will have positive thermal diffusion factors (meaning that the heavier particles will accumulate at the colder side of the system). We have speculated that if the lighter particles also have the deeper intermolecular potential well, the thermal diffusion factor will be negative [6]. Results for the L-J/s mixture are

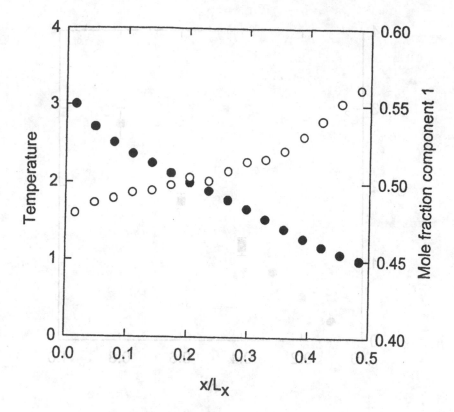

Figure 9. Profiles of temperature (●) and mole fraction of component 1 (○) in a binary L-J/s isotope mixture.

in qualitative agreement with data for binary mixtures of neutral components [6]. Although we are not yet able to predict the thermal diffusion factor for ionic systems, the numerical method has been established [33]. This will make a prediction of the salt distribution in the carbon cathode (Section 3.3) better founded in molecular theory.

During the NEMD simulations, the system develops a stationary state with temperature- and composition profiles as shown in Fig. 9 for a binary isotope mixture of L-J/s particles at a reduced number density $n^* = 0.1$, reduced temperature $T^* = 2.0$, and mass ratio $m_1/m_2 = 10$. The heat flux is preset and the mass flux is zero. With respect to the thermal diffusion factor and the heat of transfer, the main observations from our results are that both quantities depend strongly on the mass ratio as well as on the density. Fig. 10 shows that the thermal diffusion factor depends almost linearly on the density, whereas existing empirical models show a non-linear dependency [8].

The heat flux was analyzed for isotope mixtures of different densities and mass ratios by Hafskjold et al. [6]. The study showed a clear picture of how heat is transported in binary systems at low and intermediate densities; the contribution from the lighter component is predominantly flux of kinetic energy, and this contribution increases from

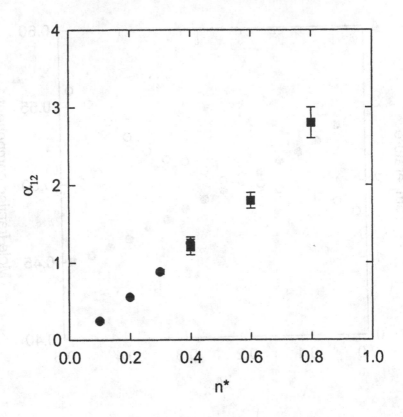

Figure 10. Thermal diffusion factor as function of density for a L-J/s binary isotope mixture. The data are for two different binary Lennard-Jones isotope mixtures; the L-J/S (●), and the switched Lennard-Jones system (■) [6]. The heat of transfer is given by $q^* = 11\alpha_{12}$ for the actual conditions of the system reported here.

the cold to the hot side. The contribution from the heavier component is predominantly intermolecular energy transfer through molecular interactions, and it increases from the hot to the cold side. This explains why there is a Soret effect and why the thermal diffusion factor is positive; *heat is conducted more effectively through the total system if the lighter component accumulates at the hot side.*

Consider next a binary mixture having constant diffusive fluxes of components 1 and 2 across an interface. The mixture is the same L-J/s isotope mixture as discussed above. In this case, component 1 diffuses from region H to region L, and component 2 in the opposite direction. Because of the mass difference between the two components, the center of mass moves relative to the interface. When we consider the system in a cell fixed (laboratory fixed) frame of reference, the interface is stationary, however. The enthalpy flux contributions from the two components are shown in Fig. 11. The most remarkable feature is that the heavy component 1 carries energy from the bulk phases to the interface, whereas component 2 carries energy from the interface to the bulk phases.

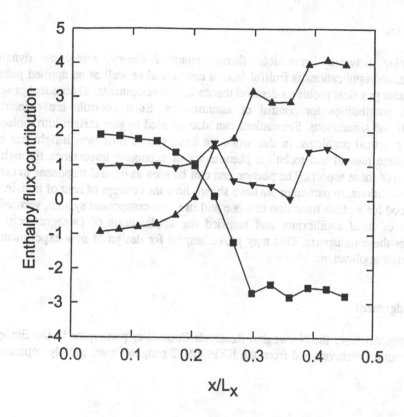

Figure 11. The enthalpy flux across an interface in a two-component L-J/s isotope mixture and with constant diffusive mass fluxes of components 1 and 2. The flux contributions are from component 1 (■) and component 2 (▲), and the total flux is represented by ▼.

This is a consequence of the fact that the dominant energy flux in the gas phase is kinetic energy flux, whereas in the liquid phase, the intermolecular energy transfer cancels the kinetic energy flux under these conditions, and the energy flux is approximately equal to the flux of potential energy. The kinetic energy flux is positive for component 1 and negative for component 2 in the left half of the NEMD cell that is shown in Fig. 11. The potential energy flux has the opposite sign of the kinetic energy flux. This feature may lead to special effects at the surface in heat transport across vapour/liquid interfaces, relevant for distillation. We shall elaborate on this in the future.

From this knowledge we predict that thermal diffusion of salts in carbon cathodes take place for the same reason. The more effective heat conduction was also used as an argument for thermal osmosis in carbon cathodes and in frost heave. This is consistent with equipartition of forces (not equipartition of the entropy production [34]), as recently pointed out by Ratkje et al. [35]. Our results for the heat transport show an entropy production per unit volume and unit time which increases from the hot to the cold side.

5. Conclusion

The interplay between irreversible thermodynamics theory, molecular dynamics simulations, and applications is fruitful from a theoretical as well as an applied point of view. Difficult practical problems demand theoretical developments. Theoretical progress depend on possibilities for control of assumptions. Such controls are offered by experiments and simulations. Simulations can also be used to gain insight into molecular aspects of practical problems. In this work we have shown how new insight has been gained in some natural and technical phenomena of economic importance through the combination of these aspects. The phenomena can be seen as natural responses to certain boundary conditions. In particular we have shown how the concept of heat of transfer can be understood for a phase transition in a one and in a two-component system, verified the assumption of local equilibrium and justified the applications of Onsager reciprocal relations for these transports. This may prove helpful for design of new experiments or new industrial applications.

Acknowledgement

Financial support from the Norwegian Research Council to participate in the European Thermodynamics Network and from the EXPOMAT program were greatly appreciated.

References

1. Førland, K.S., Førland, T. and Ratkje, S.K. (1994) *Irreversible Thermodynamics. Theory and Applications*, 2. repr. Wiley, Chichester.
2. Ratkje, S.K. and Flesland, O. (1995) *J. Food Engng.* **25**, 553-567.
3. Nygård, I. and Ratkje, S.K. (1994) *Light Metals*, Proceed. TMS Ann. Meeting, San Fransico, p.457-461.
4. Bornhorst, W.J. and Hatsopolulos, G.N. (1967) *J. Applied Mechanics* 840-?
5. Kincaid, J.M., Li, X., and Hafskjold, B. (1992) *Fluid Phase Equil.* **76**, 113-121
6. Hafskjold, B., Ikeshoji, T., and Ratkje, S.K. (1993) *Molec. Phys.* **80**, 1389-1412.
7. Ikeshoji, T. and Hafskjold, B. (1994) *Molec. Phys.* **81**, 251-261.
8. Kincaid, J.M. and Hafskjold, B. (1994) *Molec. Phys.* **82**, 1099-1114.
9. Hafskjold, B. and Ratkje, S.K. (1995) *J. Stat. Phys.* **78**, 463-494.
10. Takashi, T., Ohrai, T., Yamamoto, H., and Okamoto, J. (1980) The 2nd. Int. Symp. on Ground Freezing, Trondheim, p. 713-725.
11. Tsuneto, T. (1994) *J. Phys. Soc. Japan* **63**, 2231-2234.
12. Huige, N.J.J. (1972) *Nucleation and growth of ice crystals from water and sugar solutions in continuous stirred tank crystallizers*, Ph.D. thesis, Eindhoven University of Technology, Netherlands.
13. Heldman, D.R. (1992) Food Freezing, in *Handbook of Food Engineering*, ed. by D.R. Heldman and D.B. Lund, Marcel Dekker, New York.
14. Sørlie, M. and Øye, H.A. (1989) *Cathodes in Aluminium Electrolysis*, Aluminium Verlag, GmbH, Düsseldorf.
15. Richter, J. and Prüser, U. (1977) *Ber. Bunsenges. Phys. Chem.* **81**, 508-514.
16. Lundén, A. and Olsson, J.E. (1968) *Z. Naturforsch.* **23a**, 2045-2052.
17. Grimstvedt, A., Ratkje, S.K. and Førland, T. (1994) *J. Electrochem. Soc.* **141** (1994), 1236-1241.

18. Siljan, O.J. (1990) *Sodium aluminium fluoride attack on alumino-silicate refractories*, dr.ing. thesis, Institute of Inorganic Chemistry, The Norwegian Institute of Technology, University of Trondheim, Norway.
19. Evans, D.J. and Morriss, G.P. (1990) *Statistical mechanics of nonequilibrium liquids*. Academic Press, London.
20. Cummings, P.P and Evans, D.J. (1992) *Ind. Eng. Chem.* **31**, 1237-1252.
21. MacGowan, D. and Evans, D.J. (1986) *Phys. Rev. A* **34**, 2133-2142. See also Evans, D.J. and MacGowan, D. (1987) *Phys. Rev. A* **36**, 948-950.
22. Gillan, M.J. (1987) *J. Phys. C: Solid State Phys.* **20**, 521-538.
23. Paolini, G.V. and Ciccotti, G. (1987) *Phys. Rev. A* **35**, 5156-5166.
24. Vogelsang, R., Hoheisel, C., Paolini, G.V., and Ciccotti, G. (1987) *Phys. Rev. A* **36**, 3964-3974.
25. Holian, B.L. and Evans, D.J. (1983) *J. Chem. Phys.* **78**, 5147-5150.
26. Ashurst, W.T. and Hoover, W.G. (1975) *Phys. Rev. A* **11**, 658-678.
27. Kreuzer, H.J. (1981) *Nonequilibrium Thermodynamics and its Statistical Foundations.*, Clarendon, Oxford.
28. Tenenbaum, A., Ciccotti, G., and Gallico, R. (1982) *Phys. Rev. A* **25**, 2778-2787.
29. Haile, J.M. (1992) *Molecular dynamics simulations. Elementary methods*, John Wiley & Sons, New York.
30. Goodisman, J. (1987) *Electrochemistry: Theoretical Foundations*, Wiley - Interscience, New York.
31. Legros, J.C., Goemare, P., and Platten, J.K. (1985) *Phys. Rev. A* **32**, 1903-1905.
32. Kincaid, J.M., Cohen, E.G.D., and Lopez de Haro, M. (1987) *J. Chem. Phys.* **86**, 963-974.
33. Bresme, F., Hafskjold, B., and Wold, I. (1996) *J. Phys. Chem.* (in press)
34. Tondeur, D. and Kvaalen, E. (1987) *Ind. Eng. Chem. Res.* **26**, 56-65.
35. Ratkje, S.K., Sauar, E., Hansen, E.M., Lien, K.M., and Hafskjold, B. (1995) *Ind. Eng. Chem. Res.* **34**, 3001-3007.

LAGRANGIAN AND NETWORK FORMULATIONS OF NONLINEAR ELECTROCHEMICAL SYSTEMS

J.S. SHINER AND S. FASSARI
Physiologisches Institut, Universität Bern
Bühlplatz 5, CH-3012 Bern, Switzerland

AND

STANISLAW SIENIUTYCZ
Department of Chemical Engineering, Warsaw Technical
University, 1 Warynskiego Street, 00-645 Warsaw, Poland

Abstract. We show that the equations of motion for systems with electrochemical coupling can be derived from a dissipative Lagrangian formalism and that these systems can also be cast into an equivalent network formulation. These formulations are possible since we have shown how to cast chemical reaction kinetics into the formalisms common to electrical networks and other nonchemical systems. The examples treated are simple electrochemical cells obeying Butler-Vollmer kinetics at the electrodes and mass action kinetics for the other reactions. The chemical reactions are coupled to an electrical load and in one example to an additional electrical potential source. For the latter case we show that the stationary state is described by equations of the form of the phenomenological equations of nonequilibrium thermodynamics and that these equations display an algebraic symmetry but not the differential symmetry of Onsager reciprocity in general.

1. Introduction

Until recently the current dogma held that chemical reactions were different from other sorts of dynamic processes, at least on the phenomenological level of macroscopic chemical kinetics. The symmetry property of Onsager reciprocity [1, 2] in the phenomenological equations of near equilibrium

221

J. S. Shiner (ed.), Entropy and Entropy Generation, 221–239.
© 1996 *Kluwer Academic Publishers.*

thermodynamics is valid for many sorts of nonchemical processes without the near equilibrium restriction but is valid for systems of chemical reactions only in the near equilibrium regime, where all thermodynamic driving forces (reaction affinities) are much less than the thermal energy $k_B T$ [3, 4, 5, 6] where k_B is the Boltzmann constant and T temperature. The rates of chemical reactions are proportional to their affinities [7] only in the near equilibrium regime [3, 5], but the rates of dissipative processes in ideal mechanical and electrical systems are proportional to their conjugate driving forces essentially regardless of the distance from equilibrium. The assumption of local *thermal* equilibrium may be valid for chemical reactions whether they operate in the linear near equilibrium regime or not but is valid for transport processes only in the range of linear behavior [3, 5]. The most important difference between chemical reactions and other nonchemical processes was that they seemed to obey different formalisms on a phenomenological level. The dynamics of many sorts of dynamic processes are governed by a common formalism [8, 9]; this is evidenced for example by the analogy between the equations of motion for discrete mechanical systems and those for electrical networks. The equations of motions for chemical reactions, on the other hand, are usually expressed in terms of mass action chemical kinetics, which is apparently not consistent with the formalism common to other nonchemical processes. (There are statistical formalisms which treat both chemical and nonchemical processes [10]. However, we will not consider them further here, since we are concerned with a higher, more phenomenological level of description.)

This discrepancy between the formalism for chemical reactions and that for other, nonchemical processes leads to difficulties when treating problems with coupling between chemical reactions and nonchemical processes. One must first treat the chemical reactions with standard chemical kinetics and then treat the nonchemical processes with their appropriate formalism. Finally, one must take the chemical-nonchemical coupling into account. This last step is the difficult one. The lack of a common formalism for chemical and nonchemical processes has meant that there is no general formalism available which provides the tools for formulating the chemical-nonchemical coupling. In effect, one has had to treat each class of problems as a special case and take the chemical- nonchemical coupling into account in any way possible. In practice this has meant that in most analyses of systems with chemical-nonchemical coupling either the chemical aspects have been emphasized to the neglect of the nonchemical aspects or *vice versa*.

Recently, however, we have been able to reformulate the dynamics of chemical reactions, i.e. mass action kinetics, so that they are contained within the formalism common to other, nonchemical sorts of dynamic processes [11, 12, 13, 14, 15]. This reformulation is valid not only in the near equilibrium regime but arbitrarily far from thermodynamic equilibrium. Actually there are many different, equivalent formalisms: Lagrangian dynamics [11, 14] extended to include dissipative effects through a Rayleigh dissipation function [16], dissipative Hamiltonian dynamics [14], dissipative least action formulations [13, 14], extremum power formulations [11, 15] and network formulations [11, 17, 18]. The works cited show that chemical reactions are not fundamentally different from other sorts of dynamic processes.

Now that the same formalisms are valid for chemical reactions and non-chemical processes, it is possible to treat systems with chemical-nonchemical coupling with these formalisms in such a way that chemical and nonchemical aspects are given equal weight and the chemical-nonchemical coupling appears in a natural way as a consequence of the formalisms. Preliminary results have been presented for electro- [19] and mechanochemical [20] coupling. In this paper we treat a few simple examples of electrochemical coupling in more detail. The examples we use are based on the simple electrochemical cell of Hjelmfelt et al. [21]. The formalism we choose to use is that of Lagrangian dynamics [8, 9, 11, 14]; its concept of generalized coordinates is particularly appropriate for treating systems with chemical and nonchemical variables. As an additional result we present the generalization of our previous formulations of the mass action kinetics of uncharged species [11, 13, 14, 15] to the Butler-Vollmer kinetics of electrode reactions [10, 22], thereby showing that chemical and electrochemical kinetics share a common formalism. However, our main emphasis will be on electrochemical coupling in the sense that the chemical reactions in an electrochemical cell are coupled to the electrical load on the cell. In this sense Butler-Vollmer kinetics belongs to the chemical aspects of electrochemical coupling.

2. A Simple Electrochemical Cell

In Fig. 1 we see the simple electrochemical cell of Hjelmfelt et al. [21]. It consists of two compartments separated by a semipermeable membrane, permeable to the uncharged species X but impermeable to the charged species A^- and B^-. These species take part in the following reactions on the electrodes:

$$\begin{aligned}
electrode\ 1: &\quad A^- \leftrightarrow X + e_1^- \\
electrode\ 2: &\quad X + e_2^- \leftrightarrow B^-
\end{aligned} \tag{1}$$

where e_1^- and e_2^- are the electrons produced on the electrodes of the respective compartments. The compartments are connected to reservoirs for A^- and B^- which act as ideal sources for these species i.e. they maintain the chemical potentials of A^- and B^- in the compartments at the values of the

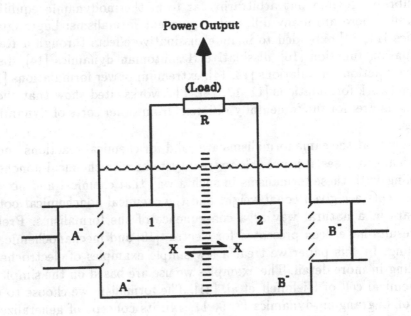

Figure 1.* The electrochemical cell treated by Hjelmfelt et al. [21].

reservoirs. The flow of X across the semipermeable membrane is assumed to occur on a much faster time scale than the two reactions above. Therefore, the flow of X from one compartment to the other can be neglected on the time scale of the reactions, and the chemical potential of X is the same in both compartments. It is further assumed that each compartment is well stirred and that the flow of counterions through the semipermeable membrane occurs on a much faster time scale than the chemical reactions, so that the electrostatic potential in both compartments is the same. Since Hjelmfelt et al. were primarily interested in the effects of the specific reaction mechanism on power production they represented the electrical load on the cell by a single equivalent resistance R. We let C_1 and C_2 be the electrical capacitances of electrodes 1 and 2, respectively, and V be the volume of the cells.

The Lagrangian formulation of dynamics for dissipative systems [8, 9, 11, 14] gives the equations of motion of a system by

$$\frac{d}{dt}\frac{\partial L}{\partial \dot{q}} - \frac{\partial L}{\partial q} + \frac{\partial \mathcal{F}}{\partial \dot{q}} + \sum_l \lambda_l \frac{\partial f_l}{\partial q} = 0 \qquad (2)$$

where L is the Lagrangian, defined as the difference between the kinetic T coenergy [9] and the potential U energy,

$$L \equiv T - U \tag{3}$$

$\mathcal{F}$ is the Rayleigh dissipation function [8, 16], q is one of the variables chosen to describe the system, which are known as generalized coordinates, t is time, and the f_l are equations of constraint, relations among the q's written so that all f_l vanish identically:

$$f_l(q) \equiv 0 \tag{4}$$

The operations indicated in eq. (2) are to be carried out with respect to each variable used to describe the systems, i.e. for each q. Actually, instead of the potential energy U in the Lagrangian of eq. (3) one should use the proper thermodynamic potential for the boundary constraints on the system. For the electrochemical cell under consideration here this is the Gibbs energy G. Therefore, we replace the Lagrangian of eq. (3) with

$$L \equiv T - G \tag{5}$$

Since inertial terms vanish on the level of macroscopic chemical kinetics the Lagrangian for the simple electrochemical cell of Fig. 1 is

$$\begin{aligned}
L = -G = & - \{ n_{A^-}\mu_{A^-} + n_{B^-}\mu_{B^-} + n_X\mu_X \\
& + \left[n_{e_1^-}\left(\mu_{e_1^-}\right) \right]_{\phi_1=0} + F^2 n_{e_1^-}^2/2C_1 \\
& + \left[n_{e_2^-}\left(\mu_{e_2^-}\right) \right]_{\phi_2=0} + F^2 n_{e_2^-}^2/2C_2 \\
& - (n_A^{ex}\mu_A^{ex} + n_B^{ex}\mu_B^{ex}) \} \\
= & - \{ n_{A^-}\mu_{A^-} + n_{B^-}\mu_{B^-} + n_X\mu_X \\
& - (q_1/F)\left(\mu_{e_1^-}\right)_{\phi_1=0} + q_1^2/2C_1 \\
& - (q_2/F)\left(\mu_{e_2^-}\right)_{\phi_2=0} + q_2^2/2C_2 \\
& - (n_A^{ex}\mu_A^{ex} + n_B^{ex}\mu_B^{ex}) \}
\end{aligned} \tag{6}$$

where the n's are mole numbers, the μ's chemical potentials, the ϕ's the electrostatic potentials at the electrodes, the superscript ex denotes the reservoirs for A^- and B^-, and the electrostatic potential of the bulk solution in the compartments is taken as the reference potential. (We are using Alberty's [23] nomenclature for the chemical potential here. What we call the chemical potential is traditionally called the electrochemical potential what is traditionally called the chemical potential is our chemical potential evaluated at the reference electrostatic potential.) F is the Faraday constant, and q_i is the charge on electrode i, related to the number of electrons by $n_{e_i^-} = q_i/F$. In both of these expressions for the Lagrangian the first line is due to the species A^-, B^- and X of the compartments of the cell, the next line represents the contribution of the electrons, and the last line is due to the sources for A^- and B^-. The contributions of the electrons to the Lagrangian have been grouped by electrode in the second line of both expressions, and those due to the electrostatic energies of the electrons at the electrodes are expressed as capacitive contributions. The last line makes a contribution of sign opposite to that of the others since we use the network convention [9] for sources. The vanishing of the inertial terms from the Lagrangian for chemical reaction systems is due to the assumption of local *thermal* equilibrium; if this assumption is relaxed we do need to include such terms [13, 14, 15]. However, we will assume that local *thermal* equilibrium is valid throughout this paper.

The dissipation function can be written in the standard quadratic form [8, 16]

$$\mathcal{F} = \frac{1}{2} \sum_j R_j \, \dot{q}_j^2 \tag{7}$$

where the $\dot{q}_j$ are the rates of the dissipative processes and the R_j their (generalized) resistances. The quadratic form arises from the observation that when the R_j do not depend on the $\dot{q}_j$ the work done by the system to overcome the resistances of the dissipative processes is $2\mathcal{F}dt$, so that the energy dissipation arising from the dissipative processes is $2\mathcal{F}$. It is often maintained that the quadratic form is valid only when the resistances of the dissipative processes are constant. However, as Rayleigh himself noted, the resistances must only be independent of the rates of the dissipative processes and may depend on the configuration of the system, i.e. on the generalized coordinates [16]. Actually even this may be too restrictive as we have recently shown; even if the resistances depend on the rates of the dissipative processes, it may be that eq. (7) is a proper form for the dissipation function and that the dissipation function is one half the energy dissipation due to the dissipative processes [15]. If we wish to maintain the quadratic form of eq. (7) in this latter case, however, the term $\partial \mathcal{F}/\partial \dot{q}$ in eq. (2) must not be taken literally as the partial derivative but simply as a short hand notation for $R_j \dot{q}_j$, the force corresponding to the dissipative process j. This need not concern us here though, since, although the resistances are in general not constant, we will write them as functions of the configuration of the system only, independent of the rates of the dissipative processes.

The dissipative processes themselves are usually straightforward to identify. In the case under consideration here we have three dissipative processes, the two reactions in eq. (1) and the flow through the resistance R of Fig. 1. As long as we write the dissipation function in terms of the rates of the dissipative processes themselves and consider one rate for each dissipative process, regardless of whether they are independent or not, then the dissipation function will have the form of eq. (7) without any cross terms. The dissipation function for the simple electrochemical cell of Hjelmfelt et al. [21] is thus

$$\mathcal{F} = \frac{1}{2} \left(R\dot{q}_R^2 + R_1^{ch}\dot{\xi}_1^2 + R_2^{ch}\dot{\xi}_2^2 \right) \tag{8}$$

where $\dot{q}_R$ is the electrical current flow through the load resistance R, and $\dot{\xi}_1$ and $\dot{\xi}_2$ are the rates of the chemical reactions (1). The R_j^{ch} are the resistances of the electrochemical reactions. When the electrostatic potentials at the electrodes vanish the resistances have the form [13, 24, 25]

$$R_j^{ch} = \frac{RT \ln \left| \left(k_j^+ \prod_{i=1}^{N} a_i^{\nu_{ij}^+} \right) / \left(k_j^- \prod_{i=1}^{N} a_i^{\nu_{ij}^-} \right) \right|}{k_j^+ \prod_{i=1}^{N} a_i^{\nu_{ij}^+} - k_j^- \prod_{i=1}^{N} a_i^{\nu_{ij}^-}} \tag{9}$$

where for reaction j k_j^+ and k_j^- are the forward and backward rate constants, respectively, ν_{ij}^+ and ν_{ij}^- are the forward and backward stoichiometric coefficients for species i, and the effect of the chemical potentials of the electrons at vanishing electrostatic potentials at the electrodes has been incorporated into the rate constants. a_i is the activity of species i, and N is the total number of chemical species in the system. For the two reactions in eq. (1) the resistances according to eq. (9) are

$$R_1^{ch} = \frac{RT \ln \left| (k_1^+ a_{A^-}) / (k_1^- a_X) \right|}{k_1^+ a_{A^-} - k_1^- a_X}$$

$$\tag{10}$$

$$R_2^{ch} = \frac{RT \ln \left| (k_2^+ a_X) / (k_2^- a_{B^-}) \right|}{k_2^+ a_X - k_2^- a_{B^-}}$$

When the electrostatic potentials at the electrodes do not vanish, their effects must be taken into account. The proper resistances then become

$$R_1^{ch} = \frac{RT \ln \left| (k_1^+ a_{A^-}) / (k_1^- a_X) \right| + Fq_1/C_1}{(k_1^+ a_{A^-}) \, e^{f_1 Fq_1/C_1 RT} - k_1^- a_X \, e^{(f_1-1)Fq_1/C_1 RT}}$$

$$\tag{11}$$

$$R_2^{ch} = \frac{RT \ln \left| (k_2^+ a_X) / (k_2^- a_{B^-}) \right| - Fq_2/C_2}{k_2^+ a_X \, e^{-f_2 Fq_2/C_2 RT} - k_2^- a_{B^-} \, e^{(1-f_2)Fq_2/C_2 RT}}$$

Here we have assumed that the reactions on the electrodes obey the Butler-Vollmer equation with symmetry constants of f_1 and f_2, respectively [10, 22]. Note that the numerators of the resistances are just the affinities

A_j of the reactions with the electrostatic energies at the electrodes taken into account:

$$A_1 = \mu_{A^-} - \left[\mu_X + \left(\mu_{e_1^-}\right)\big|_{\phi_1=0}\right] + Fq_1/C_1$$
$$A_2 = \mu_X + \left(\mu_{e_2^-}\right)\big|_{\phi_2=0} - \mu_{B^-} - Fq_2/C_2 \qquad (12)$$

All of the variables used to describe the electrochemical cell so far are not independent but are related by conservation relations for mass and charge which we write in the form of equations of constraint (4).

$$
\begin{aligned}
f_A &= n_{A^-} + \xi_1 - n_{A^-}^{ex} + \hat{c}_A = 0 \\
f_B &= n_{B^-} - \xi_2 - n_{B^-}^{ex} + \hat{c}_B = 0 \\
f_X &= n_X - \xi_1 + \xi_2 + \hat{c}_X = 0 \\
f_1 &= q_1 + F\xi_1 + q_R + \hat{c}_1 = 0 \\
f_2 &= q_2 - F\xi_2 - q_R + \hat{c}_2 = 0
\end{aligned}
\qquad (13)
$$

where the $\hat{c}$'s are integration constants, and positive current flow through the electrical resistance is taken to be from electrode 1 to electrode 2. The first three of these equations express conservation of mass for species A^-, B^- and X; the last two are just Kirchhoff's current law at the two electrodes.

Carrying out the operations indicated in eq. (2) we find the following for $q = (n_{A^-},\ n_{B^-},\ n_X,\ n_{A^-}^{ex},\ n_{B^-}^{ex},\ q_1,\ q_2,\ \xi_1,\ \xi_2,\ q_R)$:

$$
\begin{aligned}
\mu_{A^-} + \lambda_A &= 0 \\
\mu_{B^-} + \lambda_B &= 0 \\
\mu_X + \lambda_X &= 0 \\
-\mu_{A^-}^{ex} - \lambda_A &= 0 \\
-\mu_{B^-}^{ex} - \lambda_B &= 0 \\
-\left(\mu_{e_1^-}\right)_{\phi_1=0}/F + q_1/C_1 + \lambda_1 &= 0 \\
-\left(\mu_{e_2^-}\right)_{\phi_2=0}/F + q_2/C_2 + \lambda_2 &= 0 \\
R_1^{ch}\xi_1 + \lambda_A - \lambda_X + F\lambda_1 &= 0 \\
R_2^{ch}\xi_2 - \lambda_B + \lambda_X - F\lambda_2 &= 0 \\
R\dot{q}_R + \lambda_1 - \lambda_2 &= 0
\end{aligned}
\qquad (14)
$$

The first, second, third, sixth and seventh of these equations yield

$$
\begin{aligned}
\lambda_A &= -\mu_{A^-} \\
\lambda_B &= -\mu_{B^-} \\
\lambda_X &= -\mu_X \\
\lambda_1 &= -q_1/C_1 + \left(\mu_{e_1^-}\right)_{\phi_1=0}/F \\
\lambda_2 &= -q_2/C_2 + \left(\mu_{e_2^-}\right)_{\phi_2=0}/F
\end{aligned}
\qquad (15)
$$

These results plus the fourth and fifth of eqs. (14) yield

$$\mu_{A^-} = \mu_{A^-}^{ex}$$
$$\mu_{B^-} = \mu_{B^-}^{ex} \tag{16}$$

which shows that the sources (reservoirs) for species for species A^- and B^- do indeed hold the chemical potentials at the desired values.

Putting the values for the λ's from eq. (15) into the last three of eqs. (14) we find

$$R_1^{ch}\dot{\xi}_1 = \mu_{A^-} - \mu_X - \left(\mu_{e_1^-}\right)\Big|_{\phi_1 = 0} + Fq_1/C_1$$
$$R_2^{ch}\dot{\xi}_2 = \mu_X + \left(\mu_{e_2^-}\right)\Big|_{\phi_2 = 0} - \mu_{B^-} - Fq_2/C_2 \tag{17}$$
$$R\dot{q}_R = \left[q_1/C_1 - \left(\mu_{e_1^-}\right)_{\phi_1 = 0}/F\right] - \left[q_2/C_2 - \left(\mu_{e_2^-}\right)_{\phi_2 = 0}/F\right]$$

The last of these equations may also be written

$$R\dot{q}_R = \phi_1 - \phi_2 \tag{18}$$

since the chemical potential of the electrons at vanishing electrostatic potential is the same at both electrodes. This last result is just eq. (8) of Hjelmfelt et al. [21] again. The first two of eqs. (17) along with eq. (11) for the resistances of the reactions yield the kinetic equations for the reactions on the electrodes:

$$\dot{\xi}_1 = k_1^+ a_{A^-} e^{f_1 Fq_1/C_1 RT} - k_1^- a_X e^{(f_1-1)Fq_1/C_1 RT}$$
$$\dot{\xi}_2 = k_2^+ a_X e^{-f_2 Fq_2/C_2 RT} - k_2^- a_{B^-} e^{(1-f_2)Fq_2/C_2 RT} \tag{19}$$

If we replace activities with concentrations we recover the kinetic equations used by Hjelmfelt et al. [21] [their eqs. (3) and (4)].

We have previously shown the relation of the Lagrangian formulation of systems of chemical reactions to a network formulation [11, 25]. It will be instructive here to look at the network representation of the simple electrochemical cell [11, 17, 18]. The network is shown in Fig. 2. Left out of the figure are the sources for A^- and B^-. The inner part of the figure represents the electrical components of the system, the capacitances of the electrodes and the load resistance; the outer parts represent the chemical components of the system, the chemical potentials of the species A^-, B^- and X and the reactions 1 and 2. The equations of constraint (13) are just Kirchhoff's current law at the capacitors and chemical potentials. For the capacitors we have conservation of charge and for the chemical potentials, conservation of mass. The ideal transformers convert current flow into mass flow of electrons and *vice versa*. They also convert the electrical potential

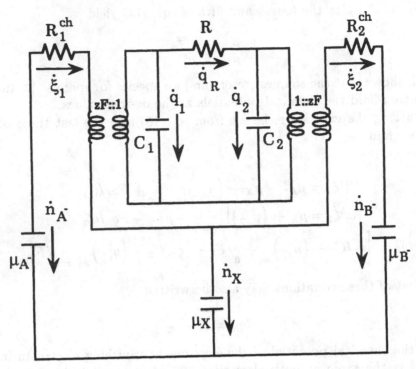

Figure 2.   The network corresponding to the electrochemical cell of Fig. 1.

differences across the capacitors into chemical potential differences in the loops for the reactions. Simple circuit analysis of the network of Fig. 2 leads immediately to the proper equations for the electrochemical cell.

3. More Complicated Reaction Schemes

Hjelmfelt et al. [21] considered not only the simple reaction scheme shown in eq. (1) but also two other, more complicated schemes. The first is

$$
\begin{aligned}
reaction\ 1: &\quad A^- \leftrightarrow X + e_1^- \\
reaction\ 2: &\quad X + 2Y \leftrightarrow 3Y \\
reaction\ 3: &\quad Y + e_2^- \leftrightarrow B^-
\end{aligned}
\tag{20}
$$

Here X can react only on electrode 1 (reaction 1), and another intermediate species Y can only react on electrode 2 to form B^- (reaction 3). The autocatalytic Schlögl reaction [26] is used for the formation of Y from X (reaction 2). The Lagrangian is

$$L = -G = -[n_{A-}\mu_{A-} + n_{B-}\mu_{B-} + n_X\mu_X + n_Y\mu_Y$$
$$- (q_1/F)\left(\mu_{e_1^-}\right)_{\phi_1=0} + q_1^2/2C_1$$
$$- (q_2/F)\left(\mu_{e_2^-}\right)_{\phi_2=0} + q_2^2/2C_2 \tag{21}$$
$$- (n_{A-}^{ex}\mu_{A-}^{ex} + n_{B-}^{ex}\mu_{B-}^{ex})]$$

The dissipation function is

$$\mathcal{F} = \frac{1}{2}\left(R\dot{q}_R^2 + R_1^{ch}\dot{\xi}_1^2 + R_2^{ch}\dot{\xi}_2^2 + R_3^{ch}\dot{\xi}_3^2\right) \tag{22}$$

where is R_1^{ch} given by eq. (11); R_3^{ch} is given by the expression for R_2^{ch} in eq. (11) with a_X replaced by a_Y, and k_2^+ and k_2^- replaced by k_3^+ and k_3^-, respectively; and R_2^{ch} is given by the general expression for purely chemical resistances [eq. (9)], since reaction 2 has no charged components:

$$R_2^{ch} = \frac{RT \ \ln\left|(k_2^+ a_X)/(k_2^- a_Y)\right|}{k_2^+ a_X a_Y^2 - k_2^- a_Y^3} \tag{23}$$

The equations of constraint are

$$f_A = n_{A-} + \xi_1 - n_{A-}^{ex} + \hat{c}_A = 0$$
$$f_B = n_{B-} - \xi_3 - n_{B-}^{ex} + \hat{c}_B = 0$$
$$f_X = n_X - \xi_1 + \xi_2 + \hat{c}_X = 0$$
$$f_Y = n_Y - \xi_2 + \xi_3 + \hat{c}_Y = 0 \tag{24}$$
$$f_1 = q_1 + F\xi_1 + q_R + \hat{c}_1 = 0$$
$$f_2 = q_2 - F\xi_3 - q_R + \hat{c}_2 = 0$$

Carrying out the operations indicated in eq. (2) we find the following for $q = (n_{A-}, \ n_{B-}, \ n_X, \ n_Y, \ n_{A-}^{ex}, \ n_{B-}^{ex}, \ q_1, \ q_2, \ \xi_1, \ \xi_2, \ \xi_3, \ q_R)$:

$$\mu_{A-} + \lambda_A = 0$$
$$\mu_{B-} + \lambda_B = 0$$
$$\mu_X + \lambda_X = 0$$
$$\mu_Y + \lambda_Y = 0$$
$$-\mu_{A-}^{ex} - \lambda_A = 0$$
$$-\mu_{B-}^{ex} - \lambda_B = 0$$
$$-\left(\mu_{e_1^-}\right)_{\phi_1=0}/F + q_1/C_1 + \lambda_1 = 0 \tag{25}$$
$$-\left(\mu_{e_2^-}\right)_{\phi_2=0}/F + q_2/C_2 + \lambda_2 = 0$$
$$R_1^{ch}\dot{\xi}_1 + \lambda_A - \lambda_X + F\lambda_1 = 0$$
$$R_2^{ch}\dot{\xi}_2 + \lambda_X - \lambda_Y = 0$$
$$R_3^{ch}\dot{\xi}_3 - \lambda_B + \lambda_Y - F\lambda_2 = 0$$
$$R\dot{q}_R + \lambda_1 - \lambda_2 = 0$$

Using the first four and the seventh and eighth of these equations to solve for the undetermined multipliers and substituting those results in the remainder of the equations, we find eq. (16)

$$\mu_{A-} = \mu_{A-}^{ex}$$
$$\mu_{B-} = \mu_{B-}^{ex}$$

again, and

$$R_1^{ch}\dot{\xi}_1 = \mu_{A-} - \mu_X - \left(\mu_{e_1^-}\right)_{\phi_1=0} + Fq_1/C_1$$
$$R_2^{ch}\dot{\xi}_2 = \mu_X - \mu_Y$$
$$R_3^{ch}\dot{\xi}_3 = \mu_Y + \left(\mu_{e_2^-}\right)_{\phi_2=0} - \mu_{B-} - Fq_2/C_2 \tag{26}$$
$$R\dot{q}_R = \left[q_1/C_1 - \left(\mu_{e_1^-}\right)_{\phi_1=0}/F\right] - \left[q_2/C_2 - \left(\mu_{e_2^-}\right)_{\phi_2=0}/F\right]$$

The last of these leads to eq. (18)

$$R\dot{q}_R = \phi_1 - \phi_2$$

again, and the first three along with the definitions of the resistances for the reactions yield the kinetic equations for the reactions:

$$\dot{\xi}_1 = k_1^+ a_{A-} e^{f_1 Fq_1/C_1 RT} - k_1^- a_X e^{(f_1-1)Fq_1/C_1 RT}$$
$$\dot{\xi}_2 = k_2^+ a_X a_Y^2 - k_2^- a_Y^3 \tag{27}$$
$$\dot{\xi}_3 = k_3^+ a_Y e^{-f_2 Fq_2/C_2 RT} - k_3^- a_{B-} e^{(1-f_2)Fq_2/C_2 RT}$$

The network representation of reactions (20) is shown in Fig. 3. Of particular interest are the ideal transformers representing the stoichiometry of species Y in reaction 2. [18]

The second more complicated reaction scheme used by Hjelmfelt et al. [21] adds the following two reactions to those of eq. (20):

$$\text{reaction 4}: \quad X \leftrightarrow Z + D$$
$$\text{reaction 5}: \quad Z \leftrightarrow Y \tag{28}$$

Species D is controlled by an ideal source in the same manner as species A^- and B^-. The Lagrangian is

$$L = -G = -[n_{A-}\mu_{A-} + n_{B-}\mu_{B-} + n_D\mu_D + n_X\mu_X + n_Y\mu_Y + n_Z\mu_Z$$
$$- (q_1/F)\left(\mu_{e_1^-}\right)_{\phi_1=0} + q_1^2/2C_1$$
$$- (q_2/F)\left(\mu_{e_2^-}\right)_{\phi_2=0} + q_2^2/2C_2$$
$$- (n_{A-}^{ex}\mu_{A-}^{ex} + n_{B-}^{ex}\mu_{B-}^{ex} + n_D^{ex}\mu_D^{ex})] \tag{29}$$

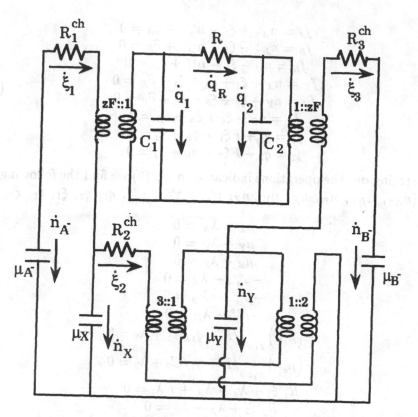

Figure 3. The network corresponding to the electrochemical cell of Fig. 1 with the reactions given by eq. (20).

The dissipation function is

$$\mathcal{F} = \frac{1}{2} \left(R\dot{q}_R^2 + R_1^{ch}\dot{\xi}_1^2 + R_2^{ch}\dot{\xi}_2^2 + R_3^{ch}\dot{\xi}_3^2 + R_4^{ch}\dot{\xi}_4^2 + R_5^{ch}\dot{\xi}_5^2 \right) \qquad (30)$$

where the two new resistances are given by [eq. (9)]:

$$R_4^{ch'} = RT \ \ln\left|(k_4^+ a_X)/(k_4^- a_Z a_D)\right|/(k_4^+ a_X - k_4^- a_Z a_D)$$

$$R_5^{ch} = RT \ \ln\left|(k_5^+ a_Z)/(k_5^- a_Y)\right|/(k_5^+ a_Z - k_5^- a_Y) \qquad (31)$$

The equations of constraint are

$$f_A = n_{A^-} + \xi_1 - n_{A^-}^{ex} + \widehat{c}_A = 0$$
$$f_B = n_{B^-} - \xi_3 - n_{B^-}^{ex} + \widehat{c}_B = 0$$
$$f_D = n_D - \xi_4 - n_D^{ex} + \widehat{c}_D = 0$$
$$f_X = n_X - \xi_1 + \xi_2 + \xi_4 + \widehat{c}_X = 0$$
$$f_Y = n_Y - \xi_2 + \xi_3 - \xi_5 + \widehat{c}_Y = 0 \tag{32}$$
$$f_Z = n_Z - \xi_4 + \xi_5 + \widehat{c}_Z = 0$$
$$f_1 = q_1 + F\xi_1 + q_R + \widehat{c}_1 = 0$$
$$f_2 = q_2 - F\xi_3 - q_R + \widehat{c}_2 = 0$$

Carrying out the operations indicated in eq. (2) we find the following for
$q = (n_{A^-}, \ n_{B^-}, \ n_D, \ n_X, \ n_Y, \ n_Z, \ n_{A^-}^{ex}, \ n_{B^-}^{ex}, \ n_D^{ex}, \ q_1, \ q_2, \ \xi_1, \ \xi_2, \ \xi_3, \ \xi_4, \ \xi_5$

$$\mu_X + \lambda_X = 0$$
$$\mu_Y + \lambda_Y = 0$$
$$\mu_Z + \lambda_Z = 0$$
$$-\mu_{A^-}^{ex} - \lambda_A = 0$$
$$-\mu_{B^-}^{ex} - \lambda_B = 0$$
$$-\mu_D^{ex} - \lambda_D = 0$$
$$-\left(\mu_{e_1^-}\right)_{\phi_1=0}/F + q_1/C_1 + \lambda_1 = 0$$
$$-\left(\mu_{e_2^-}\right)_{\phi_2=0}/F + q_2/C_2 + \lambda_2 = 0 \tag{33}$$
$$R_1^{ch}\xi_1 + \lambda_A - \lambda_X + F\lambda_1 = 0$$
$$R_2^{ch}\dot{\xi}_2 + \lambda_X - \lambda_Y = 0$$
$$R_3^{ch}\dot{\xi}_3 - \lambda_B + \lambda_Y - F\lambda_2 = 0$$
$$R_4^{ch}\dot{\xi}_4 - \lambda_D + \lambda_X - \lambda_Z = 0$$
$$R_5^{ch}\dot{\xi}_5 - \lambda_Y + \lambda_Z = 0$$
$$R\dot{q}_R + \lambda_1 - \lambda_2 = 0$$

Solving for the undetermined multipliers and substituting, we find

$$\mu_{A^-} = \mu_{A^-}^{ex}$$
$$\mu_{B^-} = \mu_{B^-}^{ex}$$
$$\mu_D = \mu_D^{ex}$$
$$R_1^{ch}\dot{\xi}_1 = \mu_{A^-} - \mu_X - \left(\mu_{e_1^-}\right)_{\phi_1=0} + Fq_1/C_1$$
$$R_2^{ch}\dot{\xi}_2 = \mu_X - \mu_Y \tag{34}$$
$$R_3^{ch}\dot{\xi}_3 = \mu_Y + \left(\mu_{e_2^-}\right)_{\phi_2=0} - \mu_{B^-} - Fq_2/C_2$$
$$R_4^{ch}\dot{\xi}_4 = \mu_X - \mu_D - \mu_Z$$
$$R_5^{ch}\dot{\xi}_5 = \mu_Z - \mu_Y$$
$$R\dot{q}_R = \left[q_1/C_1 - \left(\mu_{e_1^-}\right)_{\phi_1=0}/F\right] - \left[q_2/C_2 - \left(\mu_{e_2^-}\right)_{\phi_2=0}/F\right]$$

which yields eqs. (16), (18) and (27) and

$$\dot{\xi}_4 = k_4^+ a_X - k_4^- a_Z a_D$$
$$\dot{\xi}_5 = k_5^+ a_Z - k_5^- a_Y \tag{35}$$

for the last two reactions when the resistances for the reactions are taken into account. It may be instructive for the reader to construct the network for this case.

4. An Electrochemical Cell With a More Realistic Electrical Load

Hjelmfelt et al. [21] wished to investigate the effects of nonlinear reaction mechanisms on the electrochemical cell and therefore lumped the electrical load into a single resistor. To investigate the electrochemical coupling more fully and show how the Lagrangian formulation simplifies its treatment we will now consider a more complicated electrical load including an electrical potential source. We will use the simple reaction scheme of eq. (1) for this purpose. To the electrochemical cell of Fig. 1 we add a second resistance R_2 and the electrical potential source E. The network of Fig. 2 then becomes that of Fig. 4. Note that the current conjugate to E flows out of the positive pole since we are using the network convention. Also note that the original single electrical resistance of Fig. 1 and Fig. 2 has been renamed R_1.

The Lagrangian for this electrochemical cell connected to an external circuit with an electrical potential source is

$$
\begin{aligned}
L = -G = - [& n_{A^-} \mu_{A^-} + n_{B^-} \mu_{B^-} + n_X \mu_X \\
& - (q_1/F) \left(\mu_{e_1^-} \right)_{\phi_1 = 0} + q_1^2/2C_1 \\
& - (q_2/F) \left(\mu_{e_2^-} \right)_{\phi_2 = 0} + q_2^2/2C_2 \\
& - (n_{A^-}^{ex} \mu_{A^-}^{ex} + n_{B^-}^{ex} \mu_{B^-}^{ex} + E q_E)]
\end{aligned}
\tag{36}
$$

The dissipation function is

$$\mathcal{F} = \frac{1}{2} \left(R_1 \dot{q}_{R_1}^2 + R_2 \dot{q}_{R_2}^2 + R_1^{ch} \dot{\xi}_1^2 + R_2^{ch} \dot{\xi}_2^2 \right) \tag{37}$$

where the chemical resistances are given by eq. (11) above again. The equations of constraint are

$$
\begin{aligned}
f_A &= n_{A^-} + \xi_1 - n_{A^-}^{ex} + \hat{c}_A = 0 \\
f_B &= n_{B^-} - \xi_2 - n_{B^-}^{ex} + \hat{c}_B = 0 \\
f_X &= n_X - \xi_1 + \xi_2 + \hat{c}_X = 0 \\
f_1 &= q_1 + F\xi_1 + q_{R_1} - q_{R_2} + \hat{c}_1 = 0 \\
f_2 &= q_2 - F\xi_2 - q_{R_1} + q_E + \hat{c}_2 = 0 \\
f_E &= q_E - q_{R_2} + \hat{c}_E = 0
\end{aligned}
\tag{38}
$$

Performing the operations indicated in eq. (2) and eliminating the undetermined multipliers we find the equations of motion

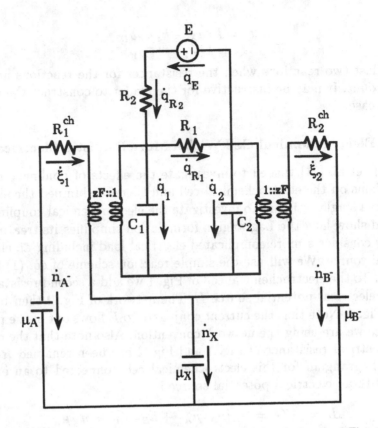

Figure 4. The network corresponding to the electrochemical cell of Fig. 1 with an external electrical potential source added.

$$R_1^{ch}\dot{\xi}_1 = \mu_{A^-} - \mu_X - \left(\mu_{e_1^-}\right)_{\phi_1=0} + Fq_1/C_1$$
$$R_2^{ch}\dot{\xi}_2 = \mu_X + \left(\mu_{e_2^-}\right)_{\phi_2=0} - \mu_{B^-} - Fq_2/C_2 \qquad (39)$$
$$R_1\dot{q}_{R_1} = q_1/C_1 - q_2/C_2$$
$$R_2\dot{q}_{R_2} = E - (q_1/C_1 - q_2/C_2)$$

At the stationary state $\dot{n}_X$, $\dot{q}_1$ and $\dot{q}_2$ must vanish identically since they are not controlled directly by any of the external sources in the system. Making use of this condition by setting the time derivatives of the appropriate equations of constraint equal to zero, using the results to eliminate μ_X, q_1 and q_2 from the equations of motion (39), we find after further manipulation that there are two independent equations at the stationary state:

$$\mu_{A^-}^{ex} - \mu_{B^-}^{ex} = (R_1^{ch} + R_2^{ch} + F^2 R_1) J_c - FR_1 J_e$$
$$E = -FR_1 J_c + (R_1 + R_2) J_e \qquad (40)$$

where $\dot{\xi}_1 = \dot{\xi}_2 = J_c$ is the rate of the chemical reactions and $q_E = \dot{q}_{R_2} = J_e$ is the current through the source E and resistance R_2 of the external circuit. Eqs. (40) show very clearly the electrochemical coupling and are of the form of the phenomenological equations of nonequilibrium thermodynamics

$$X_i = \sum_j R_{ij} J_j \tag{41}$$

since $\mu_{A-}^{ex} - \mu_{B-}^{ex}$ and E are the phenomenological forces conjugate to the chemical and electrical flows J_c and J_e, respectively. Eqs. (40) obey an algebraic symmetry property, $R_{ce} = R_{ec} = -FR_1$. The differential symmetry of Onsager reciprocity is in general not valid, however: $\partial X_c/\partial J_e \neq \partial X_e/\partial J_c$. We would not expect the reciprocity property to hold since we have not assumed that we are in the near equilibrium regime and since the chemical resistances are highly nonlinear. The electrical resistances may also be nonlinear since we have put no restrictions on them. On the other hand, the algebraic symmetry is a property of a general class of systems, of which the electrochemical cells treated here are examples [27].

In closing this section we note that the stationary state equations (40) for this simple example could have been written down by inspection of the network of Fig. 4 by eliminating the capacitances (including the one for species X).

5. Discussion

We have demonstrated in this paper that the reformulation of the dynamics of systems of chemical reactions, mass action kinetics, so that they are given by Lagrangian dynamics, not only leads to a unification of the dynamics of chemical reactions with those of other, nonchemical processes on a phenomenological level but also allows for the treatment of chemical-nonchemical coupling in such a way that chemical and nonchemical aspects are treated on an equal footing and the coupling results naturally from the formalism. Along the way we have generalized our previous results for the resistance of a chemical reaction involving uncharged species to the case of electrode reactions obeying Butler-Vollmer kinetics, thereby emphasizing that chemical and electrochemical kinetics have a common formulation. Furthermore, we have shown that the relation between the Lagrangian and network formalisms allows for a network description of electrochemical coupling which yields insight into the physics of the system. The dynamics of chemical reactions have been formulated in terms of networks previously, with either bond graphs [17] or linear networks [18], but electrochemical coupling was not considered. The results presented here are not limited to the simple examples of an electrochemical cell treated here but should have general applicability to systems with chemical-nonchemical coupling. The algebraic symmetry shown for the last example is a general property valid for a broad class of systems [27].

We have chosen to work in terms of Lagrangian dynamics in this paper, but we have previously shown that the Lagrangian formulation of chemical reaction dynamics is equivalent to various other variational and extremal formulations, including Hamiltonian dynamics of dissipative systems and dissipative least action formulations [14, 15]. In addition we have derived a new power function, the minimization of which is equivalent to the variational formulations [15]. This new minimal principle is valid for systems for which the dissipation function can be expressed in the quadratic form of eq. (7), and shows that the resistances are not restricted to be functions of the state of the system only, as seems to be the case for the Lagrangian formulation when the dissipation function has a quadratic form; the resistances may also depend on the generalized velocities or accelerations, so that the formulation covers a broad class of systems. In addition the requirement of local *thermal* equilibrium for the chemical species can easily be relaxed in all the variational and extremal formulations [13, 14, 15].

We expect these sorts of formulation to be of value in the future for the treatment of chemical-nonchemical coupling. We have formulated the electrochemical mechanism of conduction in nerve fibers [19], the action potential, as well as the mechanochemical phenomenon of muscle contraction [20] in these terms. In the latter case, we have not only found a mechanical equation of motion, which has been lacking in previous approaches; we have also discovered the possible existence of a new mechanical force due to dissipative coupling of the mechanical motion to the chemical reactions. These are just a few examples of the sort of results we can expect in the future.

Acknowledgements

This work was supported by grants no. 31-29953.90 and 21-42049.94 from the Swiss National Science Foundation and grant no. 93.0106 from the Swiss Federal Office·for Education and Science within the scope of the EU Human Capital and Mobilities project "Thermodynamics of Complex Systems" (no. CHRX-CT92- 0007).

References

1. Onsager, L. (1931) *Phys. Rev.* **37**, 405-426.
2. Onsager, L. (1931) *Phys. Rev.* **38**, 2265-2279.
3. Prigogine, I. (1967) *Introduction to thermodynamics of irreversible processes*, Wiley, New York.

4. Glansdorff, P. and Prigogine, I. (1971) *Thermodynamic Theory of Structure, Stability and Fluctuations*, Wiley, New York.
5. Nicolis, G. and Prigogine, I. (1977) *Self-Organization in Nonequilibrium Systems*, Wiley, New York.
6. Keizer, J. (1977) *BioSystems.* **8**, 219-226.
7. de Donder, T. (1928) *L'Affinite*, Gauthier-Villars, Paris.
8. Goldstein, H. (1980) *Classical Mechanics*, Addison-Wesley, Reading.
9. MacFarlane, A.G.J. (1970) *Dynamical System Models*, Harrap, London.
10. Keizer, J. (1987) *Statistical Thermodynamics of Nonequilibrium Processes*, Springer, New York.
11. Shiner, J.S. (1992) in S. Sieniutycz and P. Salamon (eds.) *Flow, Diffusion and Rate Processes*, Taylor & Francis, New York, pp. 248-282.
12. Shiner, J.S. (1990) *Biophys. J.* **57**, 194a.
13. Sieniutycz, S. (1987) *Chem. Eng. Sci.* **42**, 2697-2711.
14. Sieniutycz, S. and Shiner, J.S. (1992) *Open Sys. Information Dyn.* **1**, 149-182.
15. Sieniutycz, S. and Shiner, J.S. (1992) *Open Sys. Information Dyn.* **1**, 327-348.
16. Rayleigh, J.W.S. (1945) *The Theory of Sound*, Dover, New York.
17. Oster, G.F., Perelson, A. and Katchalsky, A. (1973) *Q. Rev. Biophys.* **6**, 1-134.
18. Wyatt, J.L. (1978) *Computer Programs in Biomedicine* **8**, 180-195.
19. Shiner, J.S., Lüscher, H.-R. and Sieniutycz, S. (1992) *Experientia* **48**, A91.
20. Shiner, J.S. (1993) *Biophys. J.* **64**, A244.
21. Hjelmfelt, A., Schreiber, I. and Ross J. (1991) *J. Phys. Chem.* **95**, 6048-6053.
22. Koryta, J. and Stulik, K. (1983) *Ion-selective Electrodes*, Cambridge University Press, Cambridge,.
23. Alberty, R.A. and Silbey, R.J. (1992) *Physical Chemistry*, Wiley, New York.
24. Grabert, H., Hänggi, P. and Oppenheim, I. (1983) *Physica* **117A**, 300-316.
25. Shiner, J.S. (1987) *J. Chem. Phys.* **87**, 1089-1094.
26. Schlögl, F. (1972) *Z. Phys.* **253**, 147-161.
27. Shiner, J.S. and Sieniutycz, S. (1996) in J.S. Shiner (ed.), *Entropy and Entropy Generation*, Kluwer, Dordrecht, this volume.

INDEX

242

Understanding Chemical Reactivity

1. Z. Slanina: *Contemporary Theory of Chemical Isomerism.* 1986
    ISBN 90-277-1707-9
2. G. Náray-Szabó, P.R. Surján, J.G. Angyán: *Applied Quantum Chemistry.* 1987
    ISBN 90-277-1901-2
3. V.I. Minkin, L.P. Olekhnovich and Yu. A. Zhdanov: *Molecular Design of Tautomeric Compounds.* 1988
    ISBN 90-277-2478-4
4. E.S. Kryachko and E.V. Ludeña: *Energy Density Functional Theory of Many-Electron Systems.* 1990
    ISBN 0-7923-0641-4
5. P.G. Mezey (ed.): *New Developments in Molecular Chirality.* 1991
    ISBN 0-7923-1021-7
6. F. Ruette (ed.): *Quantum Chemistry Approaches to Chemisorption and Heterogeneous Catalysis.* 1992
    ISBN 0-7923-1543-X
7. J.D. Simon (ed.): *Ultrafast Dynamics of Chemical Systems.* 1994
    ISBN 0-7923-2489-7
8. R. Tycko (ed.): *Nuclear Magnetic Resonance Probes of Molecular Dynamics.* 1994
    ISBN 0-7923-2795-0
9. D. Bonchev and O. Mekenyan (eds.): *Graph Theoretical Approaches to Chemical Reactivity.* 1994
    ISBN 0-7923-2837-X
10. R. Kapral and K. Showalter (eds.): *Chemical Waves and Patterns.* 1995
    ISBN 0-7923-2899-X
11. P. Talkner and P. Hänggi (eds.): *New Trends in Kramers' Reaction Rate Theory.* 1995
    ISBN 0-7923-2940-6
12. D. Ellis (ed.): *Density Functional Theory of Molecules, Clusters, and Solids.* 1995
    ISBN 0-7923-3083-8
13. S.R. Langhoff (ed.): *Quantum Mechanical Electronic Structure Calculations with Chemical Accuracy.* 1995
    ISBN 0-7923-3264-4
14. R. Carbó (ed.): *Molecular Similarity and Reactivity: From Quantum Chemical to Phenomenological Approaches.* 1995
    ISBN 0-7923-3309-8
15. B.S. Freiser (ed.): *Organometallic Ion Chemistry.* 1996
    ISBN 0-7923-3478-7
16. D. Heidrich (ed.): *The Reaction Path in Chemistry: Current Approaches and Perspectives.* 1995
    ISBN 0-7923-3589-9
17. O. Tapia and J. Bertrán (eds.): *Solvent Effects and Chemical Reactivity.* 1996
    ISBN 0-7923-3995-9
18. J.S. Shiner (ed.): *Entropy and Entropy Generation.* Fundamentals and Applications. 1996
    ISBN 0-7923-4128-7